Digital Privacy and Security

Digitalization is a big trend at the moment in the whole history of human beings. However, the movement introduces unprecedented challenges in security and privacy.

The book series on Digital Privacy and Security aims to develop and disseminate understandings of innovations, paradigms, techniques, and technologies in the contexts of digital world related research and studies. It covers security, privacy, availability, and dependability issues for digital systems and applications. It welcomes emerging technologies, such as security and privacy in artificial intelligence, digital twin, blockchain, metaverse, semantic communications, and so on.

The series serves as an essential reference source for security and privacy in the digital space. It publishes thorough and cohesive overviews on state-of-the-art topics in cyber security and privacy, as well as sophisticated techniques, original research presentations and in-depth case studies in the domain. The series also provides a single point of coverage of advanced and timely emerging topics and a forum for core concepts that may not have reached a level of maturity to warrant a comprehensive textbook. The intended audience includes students, researchers, professionals, and industrial practitioners.

The quality is assured through rigorous peer review, based on the editors' review and selection and adequate refereeing by independent experts.

Weiqi Wang • Shui Yu

Machine Unlearning: Concepts and Implementations

Springer

Weiqi Wang
School of Computer Science
University of Technology Sydney
Ultimo, NSW, Australia

Shui Yu
School of Computer Science
University of Technology Sydney
Ultimo, NSW, Australia

ISSN 2731-992X ISSN 2731-9938 (electronic)
Digital Privacy and Security
ISBN 978-981-92-1141-8 ISBN 978-981-92-1142-5 (eBook)
https://doi.org/10.1007/978-981-92-1142-5

This Springer imprint is published by the registered company Springer Nature Singapore Pte Ltd.
The registered company address is: 152 Beach Road, #21-01/04 Gateway East, Singapore 189721, Singapore

Preface

In the past two decades, we have witnessed a dramatic acceleration in artificial intelligence (AI), both in core methodological advances and in the rapid diffusion of AI applications across almost every domain of society. In recommendation, prediction, and decision support, AI systems have moved from laboratory prototypes to everyday infrastructure, reshaping how people communicate, learn, work, and access services.

At the same time, the expansion has brought security and privacy concerns into sharper focus. As AI becomes more deeply embedded in digital platforms and physical environments, it interacts with large volumes of sensitive data, exposes new interfaces for abuse, and amplifies the consequences of failures. In response, legislation and regulatory frameworks around the world have increasingly foregrounded the protection of individual privacy and data rights, pushing organisations to adopt strong safeguards, clear accountability, and transparent data governance.

Against this backdrop, machine unlearning represents a major step toward strengthening privacy protection by operationalising the individual's right to be forgotten. By enabling models to remove the influence of specified training data without full retraining, unlearning offers a practical pathway to honour deletion and withdrawal requests in modern AI pipelines. More broadly, it reframes privacy not only as limiting collection but also as enabling meaningful revocation and remediation after learning has already occurred.

We classify the existing machine unlearning literature into three categories in this book: algorithm-centric methods, application and domain-centric unlearning, and privacy and security issues. In our understanding, algorithm-centric research focuses on developing general unlearning paradigms and implementation techniques, together with principled evaluation protocols for assessing forgetting effectiveness and utility preservation. Application- and domain-centric research, by contrast, adapts unlearning to specific settings where the constraints and threat models differ substantially, such as federated learning, graph learning, and foundation generative models, including large language models (LLMs) and diffusion models. Finally, research on privacy and security issues examines the new vulnerabilities introduced by unlearning workflows and interfaces, as well as the corresponding defence strategies needed to deliver unlearning that is not only effective but also safe and trustworthy in practice.

In order to serve readers flexibly, we organise this book into four parts: algorithm-centric methods, application and domain-centric unlearning, privacy and security issues, and promising future directions. Chapters 2 and 4 introduce widely used unlearning algorithms together with key evaluation principles. Chapters 5 and 8 then shift to domain settings where unlearning faces distinct constraints, including graph learning, federated learning, and generative models such as LLMs and diffusion models. In Chap. 9, we discuss the privacy and security vulnerabilities that unlearning introduces, along with emerging defence strategies. We have to note that we only introduce the concept and very basics of the tools and methods for readers to understand the related machine unlearning research. If readers wish to continue the research using these methods, we encourage them to follow the related references. We also emphasise that our coverage is AI-focused. Cryptography-oriented perspectives (e.g., cryptographic deletion and related formal primitives) have not been included in this book. Finally, Chap. 10 offers our perspective on the near-future landscape of unlearning security and privacy, with the aim of providing a useful point of reference rather than a definitive forecast.

Machine unlearning is a new field. Our purpose is to introduce interested readers to this promising research area. It is suitable for senior undergraduate students, and it can also be used for postgraduate students or researcher as a reference for their research work. We hope this book present our readers an overview of the field, and pave a starting ground for them to further explore the uncharted land of the domain in both theoretical and application perspectives.

Due to the time and knowledge limitation, we believe there are many possible problems and errors. We humbly ask you to kindly point them to us for correction in a possible future version.

We sincerely hope you will enjoy reading the book, and let us work together to explore the promising uncharted land.

Sydney, NSW, Australia
March 2026

Weiqi Wang
Shui Yu

Acknowledgements This book is a product of our research journey on privacy and security, especially in the booming field of machine unlearning. We sincerely appreciate the colleagues and friends who offered us help, comments, suggestions and guidance on our way to this point. There are too many of them, it would be a long list if we list their names here. We want to say thank you.

We have to thank the members of our research groups at the University of Technology Sydney. PhD students and academic visitors are key players in our research, and they helped establish a solid foundation for the content of the book through numerous meetings and discussions. Some of them contributed significantly in proofreading of some chapters: Feng Wu for Chap. 2, Jianhuan Huang for Chap. 3, Wanyu Li for Chap. 4, Tao Zhang for Chap. 6, Luoyu Chen for Chap. 7, Xiaoyu Ding for Chap. 8, and Mingjian Tang for Chaps. 9 and 10.

We are also grateful to the professional staff of Springer, especially Dr. Nick Zhu, for his continuous professional support and guidance.

Last but not least, we would like to thank our families for their persistent support and love.

Competing Interests The authors have no competing interests to declare that are relevant to the content of this manuscript.

Contents

Chapter 1
Introduction of Machine Unlearning

In this chapter, we present the basic knowledge of machine unlearning and offer an overview of machine unlearning solutions, different unlearning applications, and privacy and security issues in machine unlearning. Some sections could be skipped for readers who have understood the concepts already.

Some basic information about the books is also presented at the end of the chapter for readers' reference if they want to find their interested content directly.

1.1 Machine Unlearning Paradigm

As "right to be forgotten" provisions are enacted worldwide, such as the European Union's General Data Protection Regulation (GDPR) [1], machine unlearning has emerged as a key technical mechanism for operationalizing data subjects' right to erasure [2]. A growing body of work investigates algorithmic approaches and protocols for removing the influence of specified records while preserving utility on retained data.

Let us start with the machine unlearning problem statement, the common solutions of machine unlearning, and the corresponding auditing methods.

1.1.1 Machine Unlearning Problem Statement

Modern machine learning systems are usually trained once on a large dataset and then deployed as fixed models. In practice, however, the underlying training data are not static: individuals may withdraw consent, data may be corrupted or illegally collected, and new regulations may require deletions of specific records [2–4]. Conceptually, the most reliable way to satisfy a deletion request is to retrain

W. Wang, S. Yu, *Machine Unlearning: Concepts and Implementations*,
Digital Privacy and Security, https://doi.org/10.1007/978-981-92-1142-5_1

the model from scratch on the remaining data. However, full retraining is often computationally and financially prohibitive for modern large-scale systems [2, 5]. Consequently, current machine unlearning research aims to design more efficient yet comparably effective procedures that modify a trained model so that it behaves as if the designated data had never been used for training [6, 7].

Formally, let $\mathcal{A}$ be a learning algorithm that trains on a dataset D and returns original model parameters θ, i.e., $\theta = \mathcal{A}(D)$. Suppose that, after deployment, a subset of the data $D_f \subseteq D$ is requested to be forgotten (e.g., all examples associated with a particular user, class, or time period). The remaining data are $D_r = D \setminus D_f$. The gold-standard target model in this situation is the retrained model:

$$\theta^\star = \mathcal{A}(D_r),$$

obtained by re-running the original learning procedure using only the retained data. Machine unlearning asks for an efficient algorithm $\mathcal{U}$ that takes the original model θ and the forget set D_f, and produces an updated model $\theta_u = \mathcal{U}(\theta, D_f)$ that is (approximately) equivalent to $\theta^\star$.

A useful way to formalize the unlearning requirement is to compare θ_u and $\theta^\star$ under a suitable distance or divergence measure $\rho(\cdot, \cdot)$. An unlearning procedure is said to be exact [2] if $\rho(\theta_u, \theta^\star) = 0$, that is, the distribution of models produced by $\mathcal{U}$ (over any randomness in training) is indistinguishable from that of retraining on D_r. In many realistic settings, exact equivalence is too strong or too costly to achieve. Therefore, researchers have focused on approximate machine unlearning, where one aims to guarantee that $\rho(\theta_u, \theta^\star) \leq \varepsilon$ for a given small tolerance ε, while still preserving the predictive performance of the model on the retained data distribution [5].

The problem formulation naturally induces a pipeline view of machine unlearning, as illustrated in Fig. 1.1.

- Step 0, a learning algorithm $\mathcal{A}$ is first applied to the full dataset D to produce the deployed model θ.
- Step 1, a deletion request specifies the forget set D_f, which may correspond to individual samples, all examples associated with a particular user, an entire class, or even structural components such as graph nodes or subgraphs.
- Step 2, then, we apply an unlearning procedure $\mathcal{U}$ to transform θ into an unlearned model θ_u that aims to remove the influence of D_f while preserving data utility on the retained data D_r.
- Finally, Step 3 performs audit to assess whether the resulting model θ_u behaves consistently with the ideal retrained model $\theta^\star$.

Within this pipeline, existing work instantiates a variety of unlearning algorithmic strategies and different auditing tools. We present a machine unlearning family tree with corresponding representative references in Fig. 1.2 and briefly introduce these key conceptions below, while more details and discussions will be provided in later chapters.

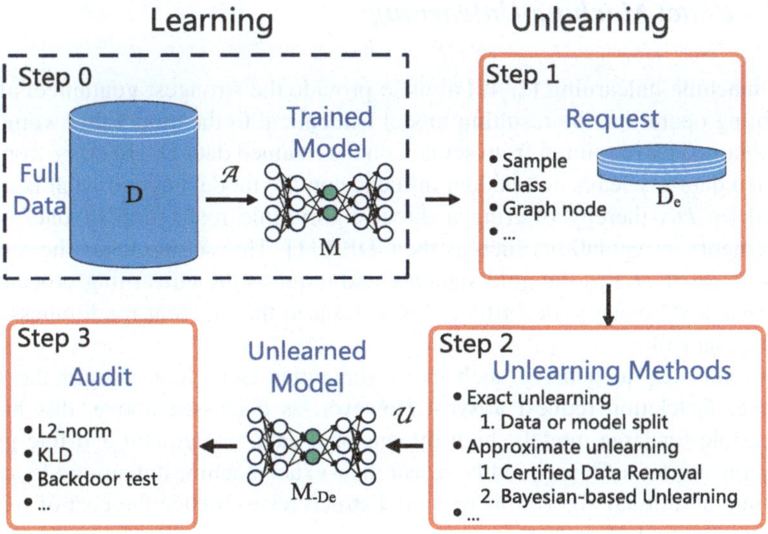

Fig. 1.1 Machine unlearning process

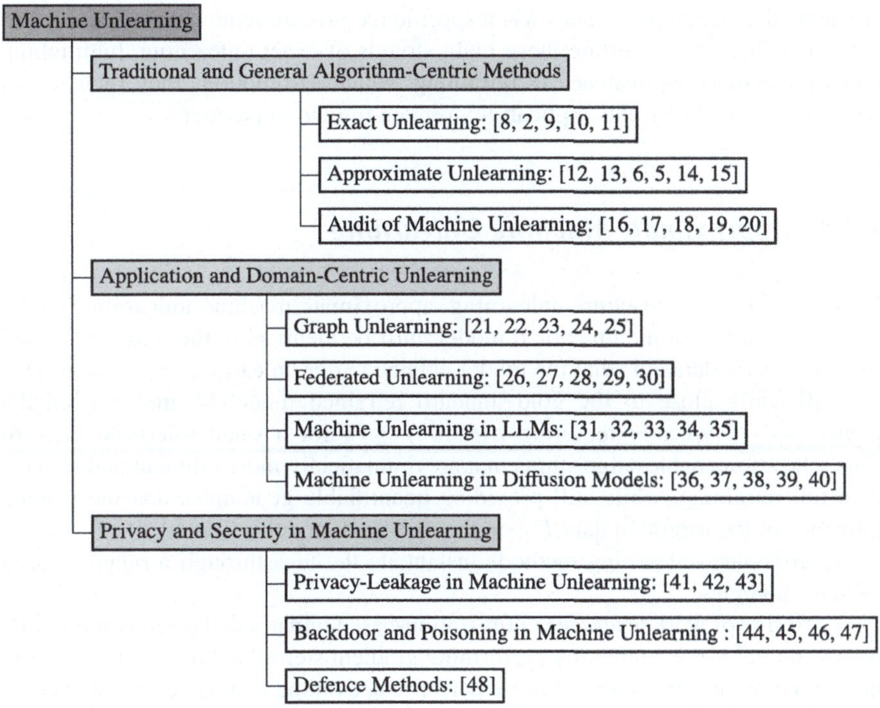

Fig. 1.2 The machine unlearning family tree

1.1.2 Exact Machine Unlearning

Exact machine unlearning [2, 49] aims to provide the strongest guarantee: after an unlearning operation, the resulting model is identical to the model that would have been obtained by retraining from scratch on the retained data D_r. In other words, the forgotten data D_f leave no residual influence on the model beyond what is already implied by D_r, thereby offering a direct algorithmic realization of data erasure requirements in regulations such as the GDPR [1]. This setting takes the retrained model $\theta^\star = \mathcal{A}(D_r)$ as the gold standard and requires the unlearning procedure $\mathcal{U}$ to reproduce $\theta^\star$ exactly, i.e., $\rho(\theta_u, \theta^\star) = 0$, up to the inherent randomness of the learning algorithm.

A trivial way to achieve such exact semantics is to fully retrain the model whenever a deletion request arrives. However, as discussed above, this baseline is infeasible for large models, high-dimensional data, or frequent and fine-grained unlearning requests. Consequently, research on exact machine unlearning focuses on exploiting additional algorithms or model structures to reduce the cost of reaching retraining optimal $\theta^\star$.

Representative approaches divide the training data or model into independently trained components that can be selectively retrained [2], and designing learning algorithms that maintain sufficient statistics or intermediate states to support efficient "decremental" updates when specific records are removed [8, 50].

In the Chap. 2, we outline these main strands of exact unlearning, highlighting how they achieve equivalence to retraining, which assumptions they rely on, and where their scalability and applicability remain limited in practice.

1.1.3 Approximate Machine Unlearning

In contrast to exact machine unlearning, approximate machine unlearning relaxes the requirement that the unlearned model must be identical to the retrained model. Instead, it only demands that the model obtained after unlearning, $\theta_u = \mathcal{U}(\theta, D_f)$, be sufficiently close to the gold-standard retrained model $\theta^\star$ under a suitable metric $\rho(\cdot, \cdot)$, typically requiring $\rho(\theta_u, \theta^\star) \leq \varepsilon$ for a small tolerance ε [5, 6]. This relaxation enables algorithms that are dramatically more efficient and scalable than full retraining, while still providing quantifiable guarantees that the residual influence of the forgotten data D_f on the deployed model is limited.

Approximate unlearning methods instantiate the idea through a range of algorithmic strategies.

A first line of work leverages optimization-based updates that directly modify the trained model, for example, by performing gradient steps that "undo" the contribution of the forget set and then restore performance on the retained data [14, 51].

A second line of work develops *certified data removal* procedures, which couple such updates with stability or convexity analyses to derive explicit upper bounds on $\rho(\theta_u, \theta^\star)$ or on the influence of any individual record [5–7].

Other approaches adopt a Bayesian perspective, interpreting unlearning as a controlled modification of the posterior that downweights or removes the likelihood contribution of D_f, and then approximates the resulting posterior distribution using variational or Monte Carlo techniques [52, 53].

These approximate methods trade the strong semantics of exact unlearning for improved practicality: they can accommodate complex models such as deep neural networks, support frequent or fine-grained deletion requests, and often integrate naturally into existing training pipelines. However, their guarantees typically depend on modeling assumptions (e.g., smooth and convex loss, well-behaved optimization trajectories), and may degrade due to distribution shift or adversarial manipulations.

Chapter 3 will examine these approximate unlearning techniques in detail, comparing their theoretical guarantees, empirical effectiveness, and robustness properties across different application domains.

1.1.4 Auditing Machine Unlearning

While exact and approximate unlearning algorithms specify what it means to forget data, in practice, one must still determine whether a particular unlearning operation has actually succeeded. Audit of machine unlearning therefore seeks to assess, using observable evidence, whether the deployed model θ_u behaves consistently with the gold-standard retrained model $\theta^\star$, and whether the residual influence of the forget set D_f has been reduced to an acceptable level [3, 19, 54]. Such audits are crucial not only for model providers, who need internal checks on their unlearning pipelines, but also for regulators and external stakeholders who may have only limited access to the system.

At a high level, existing auditing methods instantiate the abstract distance measure $\rho(\cdot, \cdot)$ in different ways. In each case, the goal is to obtain statistical evidence that the model's predictions no longer meaningfully depend on the forgotten records, while its utility on D_r remains intact.

Some approaches rely on white-box comparisons between θ_u and a reference model (often a retrained model or a suitably seeded proxy), using norms over parameters, logits, or intermediate representations, or divergences such as the Kullback–Leibler (KL) divergence [55] between output distributions on carefully chosen probe datasets [5, 19].

Besides these directly calculated distance metrics, neural network-based reconstruction methods have been proposed to analyze the difference between the unlearned and the original models to evaluate the unlearning effectiveness [20, 56].

Other approaches adopt a more behavioral perspective: they compare accuracy or loss on the forget set and the retain set, or adapt privacy attacks (e.g., membership inference, influence estimation, and backdoor activation tests) as stress tests for residual memorization of D_f [17, 54].

These auditing techniques vary in their assumptions, access requirements, and strength of guarantees: some provide provable bounds when combined with certified unlearning algorithms, whereas others offer empirical but practical diagnostics that can be applied even in black-box settings [57]. Subsequent chapters (e.g., Chap. 4) will survey these verification methods in detail, examining how they are constructed, how reliably they detect unlearning failures, and how they can be integrated into end-to-end machine unlearning workflows.

1.2 Domain-Specific Machine Unlearning

The core concepts introduced above are largely agnostic to model architecture or deployment setting: they apply whenever a learning algorithm maps training data to model parameters and we later wish to erase the influence of part of that data. In practice, modern machine learning systems operate in domain-specific environments that introduce additional constraints, which in turn shape both the formulation and the feasibility of machine unlearning in specific domains.

In this section, we focus on three such domains that have attracted particular attention in recent years: graph learning [58], federated learning [59], and large language models [60] and diffusion models [61].

1.2.1 Graph Machine Unlearning

Many real-world applications represent data as graphs, where nodes encode entities (e.g., users, items, molecules, or accounts) and edges capture their relations or interactions. Graph neural networks (GNNs) trained on such data underpin modern social recommendation, fraud and intrusion detection, knowledge graph, and molecular property prediction [58, 62]. In these settings, unlearning requests rarely concern isolated feature vectors; instead, a user may ask to remove their account, a platform may be required to erase a set of interactions, or a regulator may mandate the deletion of a sensitive subgraph [24, 63]. Graph machine unlearning therefore focuses on erasing the influence of specific nodes, edges, or subgraphs on a trained graph model, while retaining predictive performance on the remaining graph structure [21, 24].

Compared with the standard i.i.d. setting, graph unlearning is challenging by the relational inductive bias of GNNs [25]: information from a single node or edge is propagated through multiple message-passing layers and can indirectly affect distant parts of the graph. Forgetting a node thus entails not only removing its own contribution, but also correcting the representations of its neighbors and downstream nodes whose embeddings have absorbed its information [24, 25].

Existing work addresses this problem through a variety of strategies, including

- structure-aware retraining on carefully partitioned graphs,
- efficient partial retraining and caching schemes,
- and perturbation-based methods that adjust node embeddings or adjacency structure to mimic training on a pruned graph.

Chapter 5 will systematically review these graph unlearning techniques, highlighting how they balance unlearning guarantees, computational efficiency, and preservation of graph-level utility.

1.2.2 Federated Machine Unlearning

Federated learning (FL) enables multiple clients (e.g., mobile devices, organizations, or edge nodes) to collaboratively train a global model by repeatedly exchanging model updates instead of raw data. This computing paradigm is attractive in privacy–sensitive domains such as mobile text input, healthcare, and finance, where data must remain local but collective intelligence is still desired [59, 64]. However, in such settings , the right to be forgotten naturally extends to federated deployments: an individual user may request that their device's contributions be removed from the global model, an institution may withdraw from a cross-silo consortium, or regulators may require that data collected during a specific period be forgotten across all participating clients [65]. Therefore, federated unlearning seeks to erase the influence of particular clients (*client-level*) or local samples (*sample-level*) on the global FL model, while preserving model utility on the remaining federated population.

At the *client level*, the goal is to eliminate all contributions of a specific client's local dataset and updates from the global model, as if that client had never participated in training. This objective is considerably more challenging than in the centralized case: the server typically has no access to raw data and often maintains only aggregated or compressed update histories, client participation is dynamic and heterogeneous, and communication and storage constraints limit how much past information can be revisited. Early frameworks such as FedEraser reconstruct a departing client's training trajectory from stored intermediate models and gradients, and then roll back its influence on the global model without accessing the original data [30]. Subsequent work improves efficiency and privacy by designing communication-efficient protocols, synthetic-update–based methods that approximate the removed client's contribution, and cryptographic schemes that provide certified bounds on the distance between the unlearned and retrained models [65, 66].

Sample-level federated unlearning operates at a finer granularity: instead of removing influence of an entire client, the system is asked to forget individual examples or small subsets within clients' local datasets. This setting offers greater flexibility for data subjects but is even more demanding technically, as the model must exploit the rest of each client's data while eliminating the specific forgotten samples. Existing approaches include communication-efficient algorithms that adjust aggregation rules to match the effect of retraining without the target samples, unified frameworks that support unlearning at multiple granularities (from individual samples up to whole classes or clients).

Vertical federated unlearning schemes that apply gradient-based corrections to unlearn specific records while keeping the remaining features and samples intact [26–29].

Chapter 6 will delve into these federated unlearning techniques in detail, examining how they trade off unlearning guarantees, communication and computation overhead, and robustness to strategic or unreliable clients.

1.2.3 Machine Unlearning in LLMs

Large language models (LLMs) have become the backbone of contemporary AI Systems deployed in search, conversational agent, code assistant, and image generation platforms [60]. These models are typically trained on vast web-scale corpora that inevitably contain sensitive personal data, copyrighted material, and harmful or biased content [67]. As a result, LLMs can memorize and regurgitate training snippets, reproduce unsafe behaviors, or generate content that violates legal or platform-specific policies [60, 68]. In this context, machine unlearning for LLMs aims to selectively remove undesirable training records, memorized content, or behavior-inducing patterns from the model while preserving its overall utility [60].

For LLMs, unlearning requests may target specific user conversations, documents, factual associations, or capability clusters (e.g., instructions for abuse, privacy violations, or policy-breaking topics). Compared with conventional supervised models, unlearning in LLMs is complicated due to the enormous parameter scale, the diversity of downstream tasks, and highly entangled representation of knowledge.

Recent work explores gradient-based fine-tuning schemes that apply "forgetting" updates on carefully curated unlearning sets while constraining the model to remain close to its original behavior via regularization or distillation [69, 70]. Other approaches design structured parameter-space interventions and editing methods that localize and suppress specific behaviors or factual associations in a small subset of parameters. They are often coupled with evaluation frameworks that systematically probe whether the target knowledge has been removed and how much collateral damage is incurred on benign capabilities [31, 60].

Chapter 7 will provide a detailed treatment of unlearning in LLMs, comparing these techniques in terms of their unlearning guarantees, fidelity preservation, robustness, and evaluation protocols.

1.2.4 *Machine Unlearning in Diffusion Models*

Large diffusion-based generative models have become a core component of modern image generation and editing systems [60]. Trained on vast web-scale image–text corpora, they can inadvertently absorb sensitive personal information and copyrighted content present in the training data. This makes machine unlearning a critical requirement for diffusion models, enabling the targeted removal of undesirable data or concepts while preserving overall generation quality and utility [60, 61].

Diffusion models introduce a complementary set of unlearning challenges in the visual domain. Here, the goal is often to prevent the generation of certain concepts, such as particular individuals, copyrighted characters, logos, or unsafe categories, without degrading the model's ability to synthesize high-quality images in general [61]. Early "concept erasure" methods fine-tune diffusion models with negative or contrastive objectives, so that prompts involving the target concept map to safe outputs [71, 72]. Recent techniques operate at a more interpretable level, correcting the concept's representation in the latent or attention space, and adding concept-preserving regularizers so that only the targeted semantics are removed [36, 73].

Chapter 8 offers an in-depth discussion of diffusion model unlearning, examining existing techniques through the lenses of unlearning guarantees, fidelity preservation, and evaluation protocols.

1.3 Privacy and Security in Machine Unlearning

Although machine unlearning is primarily motivated by privacy regulations and the protection of data subjects, the unlearning process itself can inadvertently introduce new privacy risks. Additional computations, model updates, and logging required to support unlearning may leak information about which data have been "forgotten" and how they influenced the model, enabling refined membership inference and reconstruction attacks against both the forget set and the retained data [41, 42]. At the same time, the unlearning pipeline creates a new attack surface for adversaries who seek to undermine the integrity or availability of learning systems. Malicious clients may issue carefully crafted unlearning requests to degrade model performance on targeted subpopulations, to smuggle or preserve backdoors, or to repeatedly trigger expensive unlearning procedures as a form of denial-of-service [47, 74].

These intertwined privacy and security concerns have led to a growing line of work that both characterizes leakage and threats specific to unlearning, and proposes robust defenses that integrate cryptographic guarantees, differential privacy, certified removal, and robust training. The following subsections discuss privacy leakage, security threats, and corresponding defense techniques in detail.

1.3.1 Privacy Leakage in Machine Unlearning

Privacy leakage in machine unlearning refers to how much sensitive information about a person or their data can be exposed during unlearning. Compared with a static trained model, an unlearning-enabled system exposes additional signals: the sequence of unlearning requests, intermediate model states, and auditing statistics (e.g., loss changes, gradient norms, or influence scores). An adversary can exploit these signals to infer whether a particular sample belonged to the forget set, how strongly it influenced the model, or even to reconstruct attributes of the underlying data [41, 42]. In this sense, mechanisms that are designed to protect privacy by erasing data may paradoxically create new side channels for privacy breaches if not carefully controlled.

A natural threat model is to extend classical membership inference [41] and reconstruction attacks [17] to the unlearning setting. By comparing the behaviour of the model before and after unlearning, and by training attack models on synthetic unlearning traces, adversaries can distinguish forgotten from retained samples, detect whether a user has exercised their right to erasure [41].

Other works show that approximate unlearning and auditing procedures may amplify leakage when they rely on fine-grained gradients, influence measures, and per-example losses that correlate strongly with individual training records [5, 42]. These findings motivate the design of privacy-aware unlearning pipelines that minimise auxiliary logging, aggregate or obfuscate sensitive statistics, and integrate formal protections.

1.3.2 Security Threat in Machine Unlearning

Security threats in machine unlearning arise when an active adversary exploits the unlearning pipeline to compromise the integrity and availability. Unlike privacy leakage, where the attacker primarily infers information about training data, security attacks aim to compromise the model itself, causing targeted performance degradation, embedding or preserving hidden behaviors [41, 74]. Because unlearning algorithms typically make non-trivial parameter updates conditioned on the specified forget set, carefully crafted unlearning requests can act as a powerful lever for adversarial manipulation.

One class of threats treats unlearning as a *poisoning channel*. By strategically selecting the samples or classes to be "forgotten," an adversary can bias the effective training distribution seen by the unlearned model, thereby degrading accuracy on specific subpopulations, skewing decision boundaries, and amplifying existing fairness disparities [44, 74]. When unlearning is performed repeatedly over time (e.g., in federated or online settings), such adversarial deletion requests can accumulate, gradually steer the model toward a malicious target while each individual request appears legitimate. This attack surface becomes even more pronounced when the system automatically honors unlearning requests without strong authentication and anomaly detection mechanisms.

A second line of work demonstrates that unlearning can be abused to introduce and preserve *backdoors*. In unlearning-activated backdoor attacks, the adversary first implants a dormant backdoor during training and then designs unlearning updates that selectively erase defensive patterns while leaving the backdoor pathway intact or even strengthened [47]. Because unlearning procedures often operate on small and carefully chosen forget sets, which may lack robust regularization, they can unwittingly overfit to the requested deletions. The crafted unlearning requests can create parameter configurations in which specific triggers reliably cause misclassification, while standard evaluation metrics show only minor degradation [44]. These poisoning and backdoor threats highlight the need to analyse unlearning algorithms under adversarial conditions and to develop principled defenses.

1.3.3 Defense Technique in Machine Unlearning

To mitigate the privacy and security risks outlined above, recent work has proposed a range of defense techniques that either strengthen the unlearning algorithm itself or harden the broader training and deployment pipeline.

A first strand focuses on formal algorithmic guarantees. Certified removal mechanisms and approximate unlearning schemes provide provable bounds on how close the unlearned model is to the ideal retrained model, thereby limiting the residual influence of forgotten data and constraining what adversaries can infer [5, 6].

Differential privacy (DP) has also been leveraged to design inherently stable learners whose outputs are insensitive to the inclusion or deletion of individual records, with tight analyses relating DP parameters to the "deletion capacity" of unlearning algorithms [5, 43].

In specialized settings such as federated unlearning, cryptographic protocols (e.g., secure two-party computation and secure aggregation) have been integrated with unlearning procedures to protect historical client updates, and provide certified client removal without exposing additional raw information [65].

Beyond these theoretical guarantees, defense techniques also target concrete attack vectors against unlearning. Representation-level mechanisms such as compressive representation forgetting (CRFU) explicitly reshape intermediate features so that reconstruction of the forgotten data becomes significantly harder [48]. Other works [47, 75] advocate for robust training and verification pipelines: limiting and aggregating logged statistics, regularising unlearning updates to avoid overfitting to small forget sets, detecting anomalous, and combining unlearning with adversarial training to reduce the risk of poisoning and unlearning-activated backdoor attacks.

1.4 The Structure of the Book

The remainder and book structure is as follows.

In Chap. 2, we introduce the basic idea of exact machine unlearning and its detailed implementations. We present several mainstream exact machine unlearning methods, including statistics-based unlearning, partitioned retraining, and merge and fast ensemble methods.

In Chap. 3, we summarize technologies of approximate unlearning and corresponding detailed implementations. We introduce three kinds of representative approximate unlearning solutions: gradient-based optimization methods, model-editing methods, and representation-based methods.

In Chap. 4, we discuss how to audit and verify machine unlearning. We first clarify what machine unlearning auditing means, and then survey concrete techniques, including protocol-level verification schemes, backdoor attack-based audits, model-centric auditing methods, and benchmark-level frameworks for stress-testing unlearning algorithms at scale.

In Chap. 5, we turn to graph unlearning. We formalize what it means to forget nodes, edges, and subgraphs in graph neural networks, and then detail implementation techniques, ranging from efficient graph retraining systems to heuristic/editing-based unlearning methods that support scalable unlearning on large graphs.

In Chap. 6, we focus on federated unlearning in distributed learning environments. We first introduce the client-level federated unlearning methods, then, the class- and sample-level federated unlearning implementations. We also review privacy-aware federated unlearning frameworks that provide stronger guarantees for cross-device and cross-silo settings.

In Chap. 7, we study machine unlearning in large language models (LLMs). We first clarify what unlearning means in LLMs and then examine practical implementation details, including both parametric post-hoc and non-parametric unlearning methods.

In Chap. 8, we study machine unlearning in diffusion models. We begin by introducing machine unlearning in the generative setting, and then discuss practical implementation details, focusing on both data-centric and concept-centric unlearning methods.

In Chap. 9, we investigate privacy and security issues in machine unlearning. We outline the main threat landscape, covering privacy-leakage risks, backdoor and poisoning issues that exploit the unlearning pipeline, and defence methods with robust training to build safer unlearning systems.

In Chap. 10, we summarize the book and present our understanding of the landscape of AI security and privacy in the near future for the possible reference of our readers.

References

1. Alessandro Mantelero. The EU proposal for a general data protection regulation and the roots of the 'right to be forgotten'. *Comput. Law Secur. Rev.*, 29(3):229–235, 2013.
2. Lucas Bourtoule, Varun Chandrasekaran, Christopher A Choquette-Choo, Hengrui Jia, Adelin Travers, Baiwu Zhang, David Lie, and Nicolas Papernot. Machine unlearning. In *2021 IEEE Symposium on Security and Privacy (SP)*, pages 141–159. IEEE, 2021.
3. Thanh Tam Nguyen, Thanh Trung Huynh, Zhao Ren, Phi Le Nguyen, Alan Wee-Chung Liew, Hongzhi Yin, and Quoc Viet Hung Nguyen. A survey of machine unlearning. *ACM Transactions on Intelligent Systems and Technology*, 16(5):1–46, 2025.
4. Weiqi Wang, Zhiyi Tian, Chenhan Zhang, and Shui Yu. Machine unlearning: A comprehensive survey. *arXiv preprint arXiv:2405.07406*, 2024.
5. Ayush Sekhari, Jayadev Acharya, Gautam Kamath, and Ananda Theertha Suresh. Remember what you want to forget: Algorithms for machine unlearning. *Advances in Neural Information Processing Systems*, 34, 2021.
6. Chuan Guo, Tom Goldstein, Awni Y. Hannun, and Laurens van der Maaten. Certified data removal from machine learning models. In *Proceedings of the 37th International Conference on Machine Learning, ICML 2020, 13-18 July 2020, Virtual Event*, volume 119 of *Proceedings of Machine Learning Research*, pages 3832–3842. PMLR, 2020.
7. Eli Chien, Haoyu Wang, Ziang Chen, and Pan Li. Certified machine unlearning via noisy stochastic gradient descent. *Advances in Neural Information Processing Systems*, 37:38852–38887, 2024.
8. Yinzhi Cao and Junfeng Yang. Towards making systems forget with machine unlearning. In *2015 IEEE Symposium on Security and Privacy*, pages 463–480. IEEE, 2015.
9. Aditya Golatkar, Alessandro Achille, Avinash Ravichandran, Marzia Polito, and Stefano Soatto. Mixed-privacy forgetting in deep networks. In *Proceedings of the IEEE/CVF Conference on Computer Vision and Pattern Recognition*, pages 792–801, 2021.
10. Laura Graves, Vineel Nagisetty, and Vijay Ganesh. Amnesiac machine learning. In *Proceedings of the AAAI Conference on Artificial Intelligence*, volume 35, pages 11516–11524, 2021.
11. Sanjam Garg, Shafi Goldwasser, and Prashant Nalini Vasudevan. Formalizing data deletion in the context of the right to be forgotten. In Anne Canteaut and Yuval Ishai, editors, *Advances in Cryptology - EUROCRYPT 2020 - 39th Annual International Conference on the Theory and Applications of Cryptographic Techniques, Zagreb, Croatia, May 10-14, 2020, Proceedings, Part II*, volume 12106 of *Lecture Notes in Computer Science*, pages 373–402. Springer, 2020.
12. Yinjun Wu, Edgar Dobriban, and Susan Davidson. Deltagrad: Rapid retraining of machine learning models. In *International Conference on Machine Learning*, pages 10355–10366. PMLR, 2020.
13. Ga Wu, Masoud Hashemi, and Christopher Srinivasa. PUMA: performance unchanged model augmentation for training data removal. In *Thirty-Sixth AAAI Conference on Artificial Intelligence, AAAI 2022, Thirty-Fourth Conference on Innovative Applications of Artificial Intelligence, IAAI 2022, The Twelveth Symposium on Educational Advances in Artificial Intelligence, EAAI 2022 Virtual Event, February 22 - March 1, 2022*, pages 8675–8682. AAAI Press, 2022.
14. Seth Neel, Aaron Roth, and Saeed Sharifi-Malvajerdi. Descent-to-delete: Gradient-based methods for machine unlearning. In *Algorithmic Learning Theory*, pages 931–962. PMLR, 2021.
15. Sebastian Schelter, Stefan Grafberger, and Ted Dunning. Hedgecut: Maintaining randomised trees for low-latency machine unlearning. In *Proceedings of the 2021 International Conference on Management of Data*, pages 1545–1557, 2021.
16. Xiangshan Gao, Xingjun Ma, Jingyi Wang, Youcheng Sun, Bo Li, Shouling Ji, Peng Cheng, and Jiming Chen. Verifi: Towards verifiable federated unlearning. *IEEE Transactions on Dependable and Secure Computing*, 21(6):5720–5736, 2024.

17. Hongsheng Hu, Zoran Salcic, Gillian Dobbie, Jinjun Chen, Lichao Sun, and Xuyun Zhang. Membership inference via backdooring. In Luc De Raedt, editor, *Proceedings of the Thirty-First International Joint Conference on Artificial Intelligence, IJCAI 2022, Vienna, Austria, 23-29 July 2022*, pages 3832–3838. ijcai.org, 2022.
18. David Marco Sommer, Liwei Song, Sameer Wagh, and Prateek Mittal. Towards probabilistic verification of machine unlearning. *arXiv preprint arXiv:2003.04247*, 2020.
19. Weiqi Wang, Chenhan Zhang, Zhiyi Tian, and Shui Yu. Sms: Self-supervised model seeding for verification of machine unlearning. *IEEE Transactions on Dependable and Secure Computing*, 2025.
20. Weiqi Wang, Chenhan Zhang, Zhiyi Tian, Shui Yu, and Zhou Su. Evaluation of machine unlearning through model difference. *IEEE Transactions on Information Forensics and Security*, 2025.
21. Eli Chien, Chao Pan, and Olgica Milenkovic. Certified graph unlearning. *arXiv preprint arXiv:2206.09140*, 2022.
22. Chao Pan, Eli Chien, and Olgica Milenkovic. Unlearning graph classifiers with limited data resources. In *Proceedings of the ACM Web Conference 2023*, pages 716–726, 2023.
23. Jiahao Zhang. Graph unlearning with efficient partial retraining. In *Companion Proceedings of the ACM Web Conference 2024*, pages 1218–1221, 2024.
24. Chenhan Zhang, Weiqi Wang, Zhiyi Tian, and Shui Yu. Forgetting and remembering are both you need: Balanced graph structure unlearning. *IEEE Transactions on Information Forensics and Security*, 19:6751–6763, 2024.
25. Cheng-Long Wang, Mengdi Huai, and Di Wang. Inductive graph unlearning. In *32nd USENIX Security Symposium (USENIX Security 23)*, pages 3205–3222, 2023.
26. Weiqi Wang, Chenhan Zhang, Zhiyi Tian, and Shui Yu. Fedu: Federated unlearning via user-side influence approximation forgetting. *IEEE Transactions on Dependable and Secure Computing*, 2024.
27. Weiqi Wang, Zhiyi Tian, Chenhan Zhang, An Liu, and Shui Yu. Bfu: Bayesian federated unlearning with parameter self-sharing. In *Proceedings of the 2023 ACM Asia Conference on Computer and Communications Security*, pages 567–578, 2023.
28. Ningxin Su and Baochun Li. Asynchronous federated unlearning. In *IEEE INFOCOM 2023-IEEE conference on computer communications*, pages 1–10. IEEE, 2023.
29. Yi Liu, Lei Xu, Xingliang Yuan, Cong Wang, and Bo Li. The right to be forgotten in federated learning: An efficient realization with rapid retraining. In *IEEE INFOCOM 2022 - IEEE Conference on Computer Communications, London, United Kingdom, May 2-5, 2022*, pages 1749–1758. IEEE, 2022.
30. Gaoyang2 Liu, Xiaoqiang Ma, Yang Yang, Chen Wang, and Jiangchuan Liu. Federaser: Enabling efficient client-level data removal from federated learning models. In *2021 IEEE/ACM 29th international symposium on quality of service (IWQOS)*, pages 1–10. IEEE, 2021.
31. Lingzhi Wang, Xingshan Zeng, Jinsong Guo, Kam-Fai Wong, and Georg Gottlob. Selective forgetting: Advancing machine unlearning techniques and evaluation in language models. In *Proceedings of the AAAI Conference on Artificial Intelligence*, volume 39, pages 843–851, 2025.
32. Sijia Liu, Yuanshun Yao, Jinghan Jia, Stephen Casper, Nathalie Baracaldo, Peter Hase, Yuguang Yao, Chris Yuhao Liu, Xiaojun Xu, Hang Li, et al. Rethinking machine unlearning for large language models. *Nature Machine Intelligence*, pages 1–14, 2025.
33. Zhuoran Jin, Pengfei Cao, Chenhao Wang, Zhitao He, Hongbang Yuan, Jiachun Li, Yubo Chen, Kang Liu, and Jun Zhao. Rwku: Benchmarking real-world knowledge unlearning for large language models. *Advances in Neural Information Processing Systems*, 37:98213–98263, 2024.
34. Jiaqi Li, Qianshan Wei, Chuanyi Zhang, Guilin Qi, Miaozeng Du, Yongrui Chen, Sheng Bi, and Fan Liu. Single image unlearning: Efficient machine unlearning in multimodal large language models. *Advances in Neural Information Processing Systems*, 37:35414–35453, 2024.

35. Chris Liu, Yaxuan Wang, Jeffrey Flanigan, and Yang Liu. Large language model unlearning via embedding-corrupted prompts. *Advances in Neural Information Processing Systems*, 37:118198–118266, 2024.
36. Yongliang Wu, Shiji Zhou, Mingzhuo Yang, Lianzhe Wang, Heng Chang, Wenbo Zhu, Xinting Hu, Xiao Zhou, and Xu Yang. Unlearning concepts in diffusion model via concept domain correction and concept preserving gradient. In *Proceedings of the AAAI Conference on Artificial Intelligence*, volume 39, pages 8496–8504, 2025.
37. Yimeng Zhang, Xin Chen, Jinghan Jia, Yihua Zhang, Chongyu Fan, Jiancheng Liu, Mingyi Hong, Ke Ding, and Sijia Liu. Defensive unlearning with adversarial training for robust concept erasure in diffusion models. *Advances in neural information processing systems*, 37:36748–36776, 2024.
38. Silas Alberti, Kenan Hasanaliyev, Manav Shah, and Stefano Ermon. Data unlearning in diffusion models. *arXiv preprint arXiv:2503.01034*, 2025.
39. Yihua Zhang, Chongyu Fan, Yimeng Zhang, Yuguang Yao, Jinghan Jia, Jiancheng Liu, Gaoyuan Zhang, Gaowen Liu, Ramana Rao Kompella, Xiaoming Liu, et al. Unlearncanvas: Stylized image dataset for enhanced machine unlearning evaluation in diffusion models. *arXiv preprint arXiv:2402.11846*, 2024.
40. Jing Wu, Trung Le, Munawar Hayat, and Mehrtash Harandi. Erasing undesirable influence in diffusion models. In *Proceedings of the Computer Vision and Pattern Recognition Conference*, pages 28263–28273, 2025.
41. Min Chen, Zhikun Zhang, Tianhao Wang, Michael Backes, Mathias Humbert, and Yang Zhang. When machine unlearning jeopardizes privacy. In *Proceedings of the 2021 ACM SIGSAC Conference on Computer and Communications Security*, pages 896–911, 2021.
42. Hongsheng Hu, Shuo Wang, Tian Dong, and Minhui Xue. Learn what you want to unlearn: Unlearning inversion attacks against machine unlearning. In *2024 IEEE Symposium on Security and Privacy (SP)*, pages 3257–3275. IEEE, 2024.
43. Yiyang Huang and Clément L Canonne. Tight bounds for machine unlearning via differential privacy. *arXiv preprint arXiv:2309.00886*, 2023.
44. Yangsibo Huang, Daogao Liu, Lynn Chua, Badih Ghazi, Pritish Kamath, Ravi Kumar, Pasin Manurangsi, Milad Nasr, Amer Sinha, and Chiyuan Zhang. Unlearn and burn: Adversarial machine unlearning requests destroy model accuracy. *arXiv preprint arXiv:2410.09591*, 2024.
45. Bingguang Lu, Hongsheng Hu, Yuantian Miao, Shaleeza Sohail, Chaoxiang He, Shuo Wang, and Xiao Chen. Badfu: Backdoor federated learning through adversarial machine unlearning. *arXiv preprint arXiv:2508.15541*, 2025.
46. Jimmy Z Di, Jack Douglas, Jayadev Acharya, Gautam Kamath, and Ayush Sekhari. Hidden poison: Machine unlearning enables camouflaged poisoning attacks. In *NeurIPS ML Safety Workshop*, 2022.
47. Zirui Huang, Yunlong Mao, and Sheng Zhong. {UBA-Inf}: Unlearning activated backdoor attack with {Influence-Driven} camouflage. In *33rd USENIX Security Symposium (USENIX Security 24)*, pages 4211–4228, 2024.
48. Weiqi Wang, Chenhan Zhang, Zhiyi Tian, Shushu Liu, and Shui Yu. Crfu: Compressive representation forgetting against privacy leakage on machine unlearning. *IEEE Transactions on Dependable and Secure Computing*, 2025.
49. Haonan Yan, Xiaoguang Li, Ziyao Guo, Hui Li, Fenghua Li, and Xiaodong Lin. ARCANE: an efficient architecture for exact machine unlearning. In Luc De Raedt, editor, *Proceedings of the Thirty-First International Joint Conference on Artificial Intelligence, IJCAI 2022, Vienna, Austria, 23-29 July 2022*, pages 4006–4013. ijcai.org, 2022.
50. Somnath Basu Roy Chowdhury, Krzysztof Marcin Choromanski, Arijit Sehanobish, Kumar Avinava Dubey, and Snigdha Chaturvedi. Towards scalable exact machine unlearning using parameter-efficient fine-tuning. In *The Thirteenth International Conference on Learning Representations*, 2025.
51. Weiqi2 Wang, Chenhan Zhang, Zhiyi Tian, and Shui Yu. Machine unlearning via representation forgetting with parameter self-sharing. *IEEE Transactions on Information Forensics and Security*, 19:1099–1111, 2024.

52. Quoc Phong Nguyen, Bryan Kian Hsiang Low, and Patrick Jaillet. Variational bayesian unlearning. *Advances in Neural Information Processing Systems*, 33:16025–16036, 2020.
53. Quoc Phong Nguyen, Ryutaro Oikawa, Dinil Mon Divakaran, Mun Choon Chan, and Bryan Kian Hsiang Low. Markov chain monte carlo-based machine unlearning: Unlearning what needs to be forgotten. In *ASIA CCS '22: ACM Asia Conference on Computer and Communications Security, Nagasaki, Japan, 30 May 2022 - 3 June 2022*, pages 351–363. ACM, 2022.
54. Heng Xu, Tianqing Zhu, Lefeng Zhang, and Wanlei Zhou. Really unlearned? verifying machine unlearning via influential sample pairs. *IEEE Transactions on Dependable and Secure Computing*, 2025.
55. James M. Joyce. Kullback-leibler divergence. In *International Encyclopedia of Statistical Science*, pages 720–722. Springer, Berlin, Heidelberg, 2011.
56. Weiqi Wang, Zhiyi Tian, An Liu, and Shui Yu. Tape: Tailored posterior difference for auditing of machine unlearning. In *Proceedings of the ACM on Web Conference 2025*, pages 3061–3072, 2025.
57. Nicolas Papernot, Patrick McDaniel, Ian Goodfellow, Somesh Jha, Z Berkay Celik, and Ananthram Swami. Practical black-box attacks against machine learning. In *Proceedings of the 2017 ACM on Asia conference on computer and communications security*, pages 506–519, 2017.
58. Feng Xia, Ke Sun, Shuo Yu, Abdul Aziz, Liangtian Wan, Shirui Pan, and Huan Liu. Graph learning: A survey. *IEEE Transactions on Artificial Intelligence*, 2(2):109–127, 2021.
59. Jingxue Chen, Hang Yan, Zhiyuan Liu, Min Zhang, Hu Xiong, and Shui Yu. When federated learning meets privacy-preserving computation. *ACM Computing Surveys*, 56(12):1–36, 2024.
60. Humza Naveed, Asad Ullah Khan, Shi Qiu, Muhammad Saqib, Saeed Anwar, Muhammad Usman, Naveed Akhtar, Nick Barnes, and Ajmal Mian. A comprehensive overview of large language models. *ACM Transactions on Intelligent Systems and Technology*, 16(5):1–72, 2025.
61. Florinel-Alin Croitoru, Vlad Hondru, Radu Tudor Ionescu, and Mubarak Shah. Diffusion models in vision: A survey. *IEEE transactions on pattern analysis and machine intelligence*, 45(9):10850–10869, 2023.
62. Jiaxuan You, Tianyu Du, and Jure Leskovec. Roland: graph learning framework for dynamic graphs. In *Proceedings of the 28th ACM SIGKDD conference on knowledge discovery and data mining*, pages 2358–2366, 2022.
63. Min Chen, Zhikun Zhang, Tianhao Wang, Michael Backes, Mathias Humbert, and Yang Zhang. Graph unlearning. In *Proceedings of the 2022 ACM SIGSAC conference on computer and communications security*, pages 499–513, 2022.
64. Chen Zhang, Yu Xie, Hang Bai, Bin Yu, Weihong Li, and Yuan Gao. A survey on federated learning. *Knowledge-Based Systems*, 216:106775, 2021.
65. Ziyao Liu, Huanyi Ye, Yu Jiang, Jiyuan Shen, Jiale Guo, Ivan Tjuawinata, and Kwok-Yan Lam. Privacy-preserving federated unlearning with certified client removal. *IEEE Transactions on Information Forensics and Security*, 2025.
66. Akash Dhasade, Yaohong Ding, Song Guo, Anne-Marie Kermarrec, Martijn de Vos, and Leijie Wu. Quickdrop: Efficient federated unlearning via synthetic data generation. In *Proceedings of the 25th International Middleware Conference*, pages 266–278, 2024.
67. Myeongseob Ko, Henry Li, Zhun Wang, Jonathan Patsenker, Jiachen Tianhao Wang, Qinbin Li, Ming Jin, Dawn Song, and Ruoxi Jia. Boosting alignment for post-unlearning text-to-image generative models. *Advances in Neural Information Processing Systems*, 37:85131–85154, 2024.
68. Nicolas Carlini, Jamie Hayes, Milad Nasr, Matthew Jagielski, Vikash Sehwag, Florian Tramer, Borja Balle, Daphne Ippolito, and Eric Wallace. Extracting training data from diffusion models. In *32nd USENIX security symposium (USENIX Security 23)*, pages 5253–5270, 2023.
69. Jin Yao, Eli Chien, Minxin Du, Xinyao Niu, Tianhao Wang, Zezhou Cheng, and Xiang Yue. Machine unlearning of pre-trained large language models. *arXiv preprint arXiv:2402.15159*, 2024.

70. Zheyuan Liu, Guangyao Dou, Zhaoxuan Tan, Yijun Tian, and Meng Jiang. Towards safer large language models through machine unlearning. *arXiv preprint arXiv:2402.10058*, 2024.
71. Rohit Gandikota, Joanna Materzynska, Jaden Fiotto-Kaufman, and David Bau. Erasing concepts from diffusion models. In *Proceedings of the IEEE/CVF international conference on computer vision*, pages 2426–2436, 2023.
72. Nupur Kumari, Bingliang Zhang, Sheng-Yu Wang, Eli Shechtman, Richard Zhang, and Jun-Yan Zhu. Ablating concepts in text-to-image diffusion models. In *Proceedings of the IEEE/CVF International Conference on Computer Vision*, pages 22691–22702, 2023.
73. Bartosz Cywiński and Kamil Deja. Saeuron: Interpretable concept unlearning in diffusion models with sparse autoencoders. *arXiv preprint arXiv:2501.18052*, 2025.
74. Hongsheng Hu, Shuo Wang, Jiamin Chang, Haonan Zhong, Ruoxi Sun, Shuang Hao, Haojin Zhu, and Minhui Xue. A duty to forget, a right to be assured? exposing vulnerabilities in machine unlearning services. *arXiv preprint arXiv:2309.08230*, 2023.
75. Xinyi Sheng, Wei Bao, and Liming Ge. Robust federated unlearning. In *Proceedings of the 33rd ACM International Conference on Information and Knowledge Management*, pages 2034–2044, 2024.

Chapter 2
Exact Machine Unlearning

From the previous chapter, we have learned that machine unlearning (MU) is a critical mechanism to uphold users' right to be forgotten. One representative unlearning is exact machine unlearning, which is extended from naive retraining from scratch method. Exact unlearning methods design training procedures and model structures that allow specified samples to be removed without fully repeating the retraining process. Through the study of this chapter, readers will have a basic understanding of exact machine unlearning.

2.1 What Is Exact Machine Unlearning

The purpose of exact machine unlearning is to remove the influence of specified training data from a learned model in a way that is provably equivalent to having never used those data in training. Formally, given a learning algorithm, a fixed training pipeline (e.g., preprocessing, optimization schedule), and a deletion request, exact unlearning aims to output the same model one would get by retraining that pipeline from scratch on the dataset with the requested data removed.

To provide a clearer view of the design, we group existing exact machine unlearning methods into three broad families according to how they achieve deletion guarantees.

- As summarised in Table 2.1, algebraic/statistics-based learners exploit sufficient statistics or closed-form (decremental) updates to recompute model parameters exactly after sample removal [1, 2];
- partitioned retraining schemes localise retraining to independent shards or modules while preserving equivalence to full retraining on the retained set [3, 4];
- and modular/ensemble-based methods represent training as mergeable parameter deltas or task vectors whose contributions can be algebraically added or cancelled to forget specific data [5, 6].

W. Wang, S. Yu, *Machine Unlearning: Concepts and Implementations*,
Digital Privacy and Security, https://doi.org/10.1007/978-981-92-1142-5_2

Table 2.1 Taxonomy of exact machine unlearning methods

Method type	Main idea	Representative works
Algebraic/statistics-based learners	Training admits sufficient statistics, closed-form solutions, and decremental update formulas, so that model parameters can be recomputed exactly for sample removal without full retraining.	Decomposable-statistics unlearning [1] and data removal-enabled (DaRE) forests [2]
Partitioned retraining	Training data and parameters are divided into independent shards and modules that are trained separately and later aggregated; deleting data only retrains affected parts while yielding the same model as retraining on the retained set.	SISA [3] and ARCANE [4]
Modular/ensemble-based methods	Model updates are represented as mergeable parameter deltas or task vectors, and the contribution of deleted data is cancelled by algebraic subtraction or reweighting in parameter space.	SIFT-Masks [5] and SAFE [6]

Across these categories, exactness hinges on a deterministic, reproducible pipeline and on storing minimal auxiliary states (e.g., statistics and shard checkpoints). The main trade-offs concern storage overhead, deletion latency, and utility preservation relative to full retraining. We present the details of the implementations of exact unlearning as follows.

2.2 Exact Unlearning Implementation

2.2.1 Statistics-based Unlearning

Decomposable-Statistics Unlearning Cao et al. [1] introduced a general strategy for exact and efficient machine unlearning by rewriting learning algorithms into a summation form, so that forgetting a sample amounts to updating a small set of stored statistics rather than retraining from scratch. In this framework, a learning algorithm is expressed in terms of a finite number of aggregating summations over transformed training examples, and the model depends only on these aggregates. When a user requests deletion, the system subtracts the contribution of the forgotten sample from each affected summation and recomputes the model, thereby achieving completeness (equivalence to retraining) with an asymptotic speedup proportional to the size of the training set. An example is illustrated in Fig. 2.1.

The first component of their approach focuses on non-adaptive Statistical Query (SQ) learners [7]. The key observation is that many common algorithms can be expressed as a fixed set of aggregate sums over the training set, where each term is an efficiently computable function of a labeled example [8]. Concretely, a non-

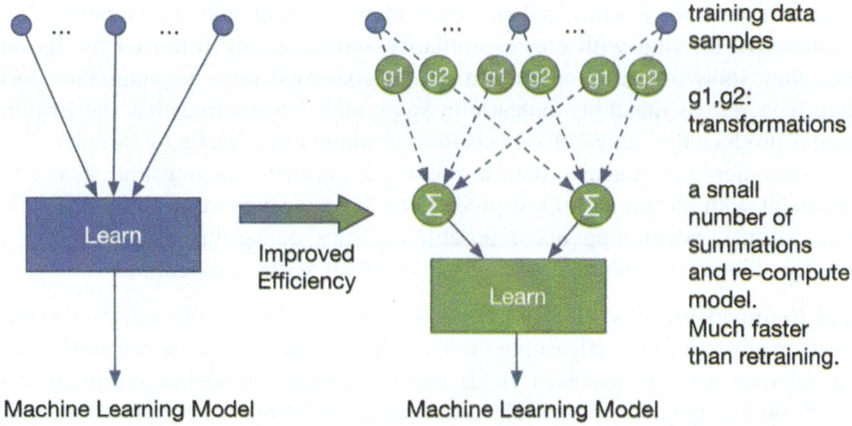

Fig. 2.1 Illustration of decomposable-statistics unlearning. **Left:** In a standard pipeline, the learned model depends directly on all training examples, so forgetting requires retraining on the retained dataset. **Right:** The learning algorithm is rewritten to depend only on a small set of summations over transformed data samples $g_i(x, l_x)$. To forget a sample, the system simply updates the corresponding summations by subtracting its contributions and then recomputes the model from these updated aggregates, which is asymptotically much faster than full retraining [1]

adaptive learner that issues m SQs is written as

$$\text{Learn}\left(\sum_{x \in X} g_1(x, l_x), \ldots, \sum_{x \in X} g_m(x, l_x)\right), \tag{2.1}$$

where each g_k is fixed in advance. Unlearning a point x_p then updates every stored summation G_k to $G_k - g_k(x_p, l_{x_p})$ and reruns the $\mathsf{Learn}(\cdot)$ step on these adjusted statistics. Cao et al. illustrate this pattern for algorithms such as naïve Bayes and Chi-squared feature selection, where probabilities and test statistics can be written as ratios of simple counts (e.g., class–feature co-occurrence counts). Hence, forgetting a sample only requires decrementing a handful of counters and recomputing the derived probabilities, yielding models that are provably identical to retraining on $X \setminus \{x_p\}$.

The second component of their approach extends decomposable-statistics unlearning to adaptive SQ and iterative algorithms, where later queries depend on earlier results and training is commonly formulated as an optimization procedure. For such learners (e.g., gradient-based methods, k-means, one-class SVM), the system stores not only the relevant summations but also the final converged state S of the algorithm. When a sample is removed, it first adjusts the summations to subtract that sample's contribution, obtaining an updated state S', and then "resumes" the iterative algorithm from S' until convergence. Under this design, completeness is guaranteed whenever the algorithm has a unique fixed point or when the resumed state is among the admissible initial states.

The third important aspect of [1] is that decomposable-statistics unlearning is applied uniformly across learning stages, covering both feature selection and model training in realistic systems. In their case studies on production-style pipelines, e.g., collaborative filtering with cosine similarity and clustering followed by decision trees, they show how each algorithm can be converted into the summation form. They then derive matching unlearning steps and demonstrate that the resulting updates produce the same outcome as full retraining after deleting a sample.

Across these components, their evaluating experiments demonstrate near 100% agreement with retraining on test predictions and speedups of up to several orders of magnitude, positioning decomposable-statistics unlearning as a foundational paradigm for exact machine unlearning in classical learning pipelines.

Data Removal-Enabled (DaRE) Forests DaRE forests modify standard random forests to support exact unlearning via localized retraining and cached statistics, so that deleting training instances yields exactly the same model as retraining from scratch on the updated dataset [2]. In line with the definition of perfect unlearning based on distributional equivalence between retrained and updated models, DaRE forests provide a certified removal guarantee with no difference between unlearned and retrained models while avoiding the prohibitive cost of full retraining after each deletion request. We present the main idea of DaRE in Fig. 2.2.

The first key idea in DaRE is to retrain only minimal subtrees by caching sufficient statistics and instance lists at each node. During initial training, each decision node stores counts of total and positive examples, along with precomputed split statistics for a set of candidate thresholds, while each leaf stores its label counts and a pointer list to all training instances that reach it. When an instance is removed, DaRE updates these cached statistics along the path from the root to the affected leaf and checks whether the locally optimal split has changed. Only if a better split emerges does it rebuild the corresponding subtree using the instances collected from its descendant leaves. Otherwise, the deletion is handled by simple count updates.

The second idea is to approximate the split search by sampling a small set of valid thresholds for each attribute, which substantially reduces both storage

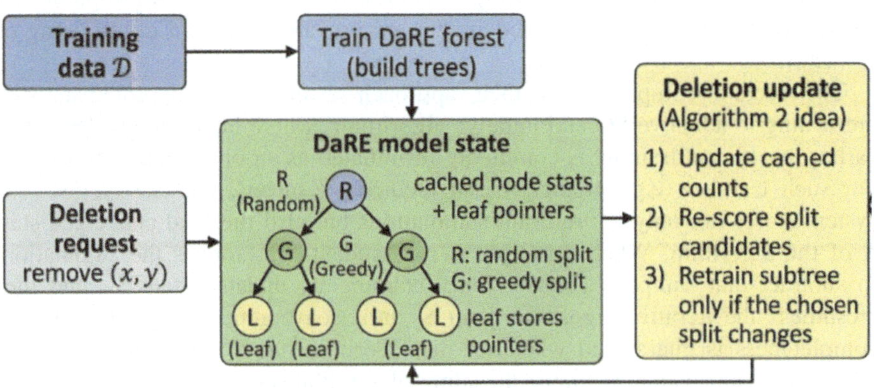

Fig. 2.2 Main idea of Data Removal-Enabled (DaRE) forests for exact unlearning

and recomputation costs while preserving predictive performance. Rather than considering all possible split points, DaRE identifies "valid" thresholds: those lying between adjacent feature values with opposite labels. Then, DaRE samples at most k such thresholds per attribute. These sampled thresholds are the only candidates whose split scores are cached and updated, so the cost of maintaining statistics scales with k rather than with the full number of potential splits. Empirical results show that modest values of k (e.g., 5–25) typically retain accuracy comparable to a standard random forest.

The third idea is to introduce random decision nodes near the top of each tree, which greatly reduces structural changes under deletions because these nodes depend only weakly on individual samples. For a random node, DaRE samples an attribute and threshold uniformly from their valid ranges and updates only simple count statistics. Retraining is triggered only when the sampled threshold becomes invalid, for example, when all instances move to one side of the split. Since random nodes are placed in the upper layers and are largely insensitive to small changes in the data distribution, they rarely need to be rebuilt, effectively isolating most deletions to lower-level greedy nodes.

Together, these components yield a random-forest variant that supports certified exact unlearning with practical efficiency across sequences of deletions. DaRE forests disable bootstrapping to simplify instance bookkeeping, train each tree independently with a random subset of attributes at each split, and expose the number of random-node layers and the number of sampled thresholds as hyperparameters that trade off update time against accuracy.

2.2.2 *Partitioned Retraining*

Since most exact unlearning methods attempted to partition the training dataset into multiple subsets and divide the ML model learning process [3, 4], we call this kind of unlearning technique partitioned retraining. The main procedure of partitioned retraining is illustrated in Fig. 2.3b.

By contrast, we first introduce the process of the naive retraining method, as shown in Fig. 2.3a. First, delete the samples from the training dataset. Second, retrain a new model using the remaining dataset. Since it needs to store the whole training dataset and retrain from the remaining dataset, it often entails significant computational and stored overhead.

The partitioned retraining based methods are proposed to reduce the retraining cost of computation and storage. As shown in Fig. 2.3b, the partitioned retraining technique can be summarized into four steps. Unlike naive unlearning trained based on the remaining trainset, the first phase of split unlearning is dividing the original full trainset into multiple disjoint shards. All the constituent models are trained based on each split data slice. Then, in the second phase, when the unlearning request comes, they only need to erase the requested samples from the split slice and retrain this slice's constituent model in the third phase. In the last phase, the

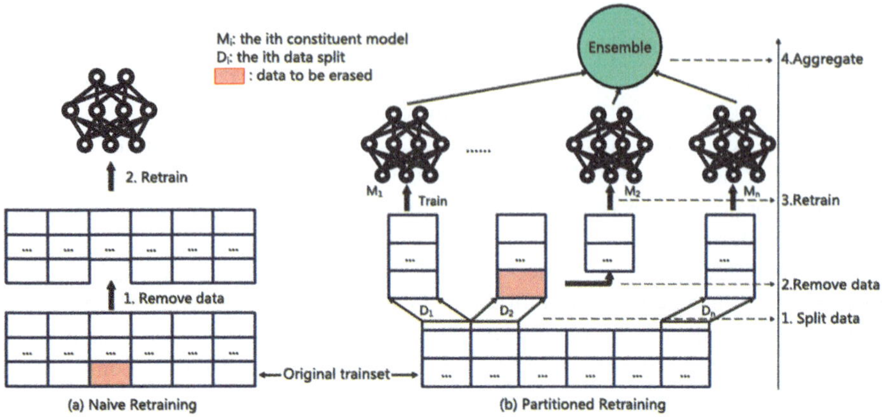

Fig. 2.3 (**a**) Naive retraining. There are only two steps: delete the specified samples from the whole dataset and retrain a model based on the remaining dataset. (**b**) Partitioned retraining. It contains four steps: (1) split the original dataset into n shards; (2) remove the erased data from the corresponding shard; (3) retrain the sub-model of this shard; and (4) ensemble all sub-models as the final model

split unlearning aggregates the retrained and other constituent models together as a new unlearned model.

Sharded, Isolated, Sliced, and Aggregated Training (SISA) One of the most representative partitioned retraining methods is SISA [3]. SISA operationalises the idea in Fig. 2.3b by carefully restructuring the original training process. First, the full training set is *sharded* into multiple disjoint subsets, and an independent constituent model is trained on each shard so that every sample only affects a single sub-model. These shard-level training runs are *isolated* from one another, which localises the impact of any future deletion request to the corresponding shard.

Within each shard, SISA further *slices* the data into an ordered sequence of mini-batches and stores checkpoints before each slice is used for training. When an unlearning request arrives, the system locates the shard and slice that contain the data to be erased, rolls back to the last checkpoint taken before that slice, deletes the requested samples, and retrains only on the remaining suffix slices. Finally, all updated shard models are *aggregated* at inference time (for example, via majority voting) to form the final predictor. In this way, SISA achieves exact unlearning in its formal model while substantially reducing the amount of retraining needed compared with naive retraining on the whole dataset.

ARCANE Yan et al. [4] adopts a similar partitioned-retraining spirit but redesigns the architecture to remain efficient even under a large number of unlearning requests, named ARCANE. Built on ensemble learning, ARCANE decomposes naive retraining into multiple one-class classification tasks, i.e., the training data are divided into several sub-datasets, and each sub-dataset is used to train an independent sub-model. When some samples must be forgotten, only the sub-models whose sub-datasets contain these samples are retrained, while the rest of the ensemble remains unchanged.

To further cut retraining time, ARCANE uses lightweight preprocessing steps. It (i) selects representative data to remove redundant samples, (ii) saves training states so retraining can resume from checkpoints instead of starting over, and (iii) orders the data so samples that are more likely to be erased appear in positions that reduce the amount of retraining needed later. Together, these design choices enable ARCANE to provide exact unlearning with substantially lower computational overhead than naive retraining and competitive accuracy compared with other partitioned retraining schemes.

RecEraser RecEraser [9] extends the partitioned retraining paradigm to recommendation by designing shard partitions and aggregation schemes that preserve collaborative information while enabling efficient exact unlearning. Concretely, RecEraser follows the SISA-style idea of splitting the interaction data into multiple shards and training an independent sub-model on each shard, but its key novelty lies in three balanced partition strategies (user-based, item-based, and interaction-based) that group similar users/items/interactions while keeping shard sizes comparable so that both collaborative signals and unlearning efficiency are maintained.

At inference time, RecEraser further introduces an attention-based adaptive aggregation mechanism that assigns instance-specific weights to shard models according to how well they capture a particular user-item pair, rather than using static or uniform voting. This lets RecEraser achieve exact recommendation unlearning by retraining only the affected shard models and the lightweight aggregation module. Meanwhile, RecEraser retains recommendation quality close to naive retraining and better than other partitioned unlearning baselines.

2.2.3 Merge and Fast Ensemble

Besides dataset partitioned retraining methods (e.g., SISA-style sharding and slicing), a growing line of work instead redesigns the training pipeline so that exact unlearning can be enacted without retraining on the critical path. The central idea is to isolate where and how each data slice influences the parameters, so that when a deletion request arrives, the unlearning server can disable only the parts of the model touched by that slice while continuing to serve with the rest. The difference between fast ensembling methods and partitioned retraining methods is that here the isolation and retraining are designed for different parameter layers of the model rather than sub-datasets and sub-models. One representative technique is Sequence-aware Sharded Sliced Training, abbreviated S^3T [10] and Synergy-Aware Forgetting Ensemble [6].

Sequence-aware Sharded Sliced Training (S^3T) S^3T extends partitioned retraining to large models that are fine-tuned on private data using parameter-efficient fine-tuning (PEFT) techniques such as LoRA [11]. The method assumes a pre-

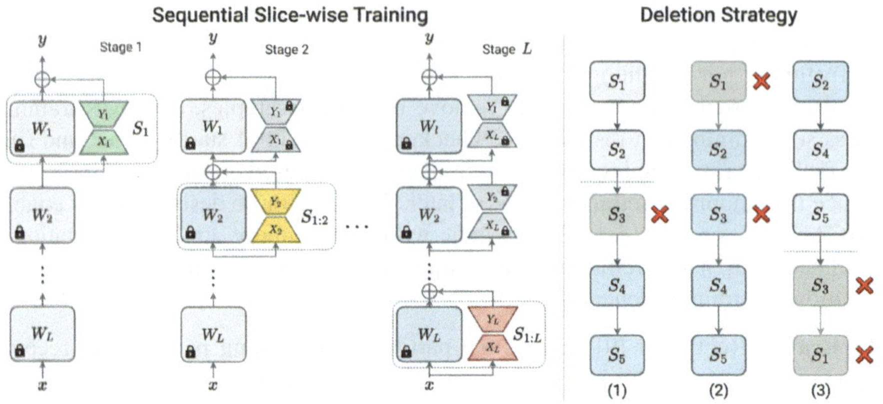

Fig. 2.4 Sequence-aware sharded sliced training in S^{3}T [10]. **Left:** Sequential slice-wise training: at stage i, only the i-th LoRA layer is updated using slices $S_{1:i}$, while all other layers remain frozen. **Right:** Deletion behaviour under different permutations of slices: removing a slice invalidates all subsequent slices in that sequence (marked with crosses), so training on diverse slice orders increases the chance that at least one model remains largely unaffected

trained backbone (e.g., BERT [12] or LLaMA [13]) that is kept fixed, while the private task-specific knowledge is injected only through lightweight PEFT adapters. The fine-tuning dataset is first split into multiple shards, and each shard is further divided into several slices. As in SISA-style schemes, a separate model is trained per shard, and the final prediction is obtained by ensembling shard-level models, but S^{3}T modifies how slices are used to train the PEFT parameters so that later unlearning can be implemented by simple layer deactivation instead of costly retraining.

The core technical idea of S^{3}T is a sequential slice-wise training strategy that achieves parameter isolation at the level of PEFT layers, as illustrated in Fig. 2.4 (left). Concretely, Chowdhury et al. [10] instantiate PEFT with LoRA and attach low-rank adapters to selected transformer layers while freezing the original weights. During training, LoRA layers are optimized in a top–down fashion, where the i-th trainable layer is updated only using the prefix of slices $\{S_1, \ldots, S_i\}$ and all other LoRA layers are either frozen or switched off. This staged schedule ensures that the LoRA parameters in deeper layers depend on progressively larger prefixes of the slice sequence. As a result, if a deletion request targets slice S_i, exact unlearning can be realized by zeroing out (deactivating) the LoRA parameters associated with the layers that have been influenced by slices $S_i, \ldots, S_L$, while keeping the pre-trained backbone and earlier slice-specific layers intact.

Beyond layer-wise isolation, S^{3}T further enhances deletion capacity by training multiple models per shard on different permutations of the slice order, as shown in Fig. 2.4 (right). Training on diverse slice sequences ensures that, for any given pattern of future deletions, there is a high probability that at least one model within each shard retains a sufficiently long unaffected prefix of slices and therefore maintains good task performance. However, enumerating all $L!$ permutations is infeasible, so S^3T introduces sequence-selection strategies under a fixed training budget. For the case of roughly uniform deletion probabilities, Chowdhury et al. [10] use an iterative cyclic-rotation algorithm to generate a small yet diverse family of permutations. When slice-wise deletion priors are non-uniform, a bipartite-matching based selection (BMS) algorithm is proposed to score candidate sequences by the expected number of surviving slices and uses a Hungarian-style matching procedure to pick a high-diversity, high-score subset.

The deletion procedure in S^3T operates purely by model selection and PEFT-layer deactivation, avoiding immediate retraining in most cases. Given a deletion request, the system locates the shard and slice that contain the offending instance, and for every model trained on that shard, it deactivates the LoRA layers associated with the affected slice positions. Models whose all LoRA layers are deactivated are dropped, and the ensemble is updated using the remaining active models. Only when too many slices across all sequences have been hit does S^3T require offline retraining, which can be scheduled without bringing the service down. Empirically, S^3T achieves accuracy comparable to or better than full fine-tuning across vision, text, and LLM benchmarks while significantly improving deletion rate and reducing total deletion time compared with SISA [3] and ProtoSISA [6], thereby providing a scalable exact-unlearning solution for large PEFT-based systems.

SAFE (Synergy-Aware Forgetting Ensemble) SAFE [6] addresses a key limitation of shard-based exact unlearning methods such as SISA, i.e., the accuracy degradation caused by training completely independent models on many small shards. The central idea is to recover part of the lost synergistic information across shards while still keeping the expected retraining cost low enough to support efficient forgetting. To this end, SAFE introduces a structured way to let shards "see" data from other shards during training, and shows that with a suitable design, one can significantly improve accuracy at a fixed forgetting budget.

The core in SAFE is the shard graph, a directed graph whose nodes represent data shards and whose edges encode which shards are allowed to share data during training. In this formalism, each node corresponds to a subset of the training data, and a directed edge (S_i, S_j) indicates that, when training the adapter associated with shard S_i, the model can also use the data stored in shard S_j. This graph structure defines both the training regime and the forgetting behaviour: more edges increase the amount of synergistic information available to each adapter, but they also enlarge the set of parameters that must be revisited when a sample in a given shard is forgotten. SAFE thus formulates the design of the shard graph as a trade-off problem between accuracy (via connectivity) and expected forgetting cost (via the size of the retraining region induced by edges).

On top of this shard graph, SAFE defines an exact forgetting procedure and analyses how different graph topologies trade off accuracy and unlearning cost [6]. In their formulation, applying a forgetting operator to a model trained on graph G should yield the same distribution as retraining from scratch on a reduced graph G' with the erased data removed. Under this definition, SISA [3] corresponds to a degenerate graph with no edges, where forgetting a sample in shard S_i only requires retraining the model for that shard. SAFE shows that naive dense and random connectivity can make the expected forgetting cost grow quickly with the node degree. Whereas, carefully structured graphs that decompose into disjoint cliques achieve a more favourable, roughly linear scaling in clique size, leading to a better accuracy–cost balance.

To better exploit label structure and support fine-grained sharding, SAFE refines the shard graph into class-specific nodes and implements shard-level classifiers with lightweight adapters on top of a frozen vision backbone [6]. Each node stores positives from a single class and draws negatives from its graph neighbours, so the training distribution for each adapter is governed by the shard graph. Using cross-attention-style adapters and pre-computed embeddings allows hundreds of such shard-specific classifiers to be trained and retrained in parallel, without the expense of training a separate deep network per shard.

SAFE further combines this graph formalism with a bilevel sharding strategy that constructs shard graphs explicitly for a given forgetting budget [6]. Coarse, class-balanced shards are first formed, and then each class within a coarse shard is partitioned into smaller "cliques" of fixed size. By tuning the clique granularity, SAFE interpolates between uniform sharding (low synergy, low cost) and highly specialised shards (high synergy but narrow coverage), and empirically finds that intermediate settings significantly improve accuracy at similar unlearning cost.

2.3 Summary

In this chapter, we introduced the notion of *exact* machine unlearning as the strongest form of deletion guarantee, where the updated model must be distributionally equivalent to retraining the same pipeline on the pruned dataset. We outlined a high-level taxonomy of existing techniques into three families: algebraic/statistics-based learners, partitioned retraining schemes, and modular or ensemble-based methods. This taxonomy provides a conceptual map of how different approaches redesign learning pipelines so that the influence of specific training samples can be cleanly removed. We then discussed representative techniques within each family.

We focused on the core ideas rather than every technical detail. For deeper understanding and practical variations, we encourage readers to consult the original papers cited in this chapter, especially those tailored to specific model classes.

Many theoretical and practical questions remain open, including tighter characterisations of exactness under randomness, better cost–utility trade-offs at scale, and extensions to federated learning, large language models, and diffusion models.

Readers who wish to explore the theoretical frontiers of machine unlearning (including its connections to stability, information theory, and privacy-preserving learning) are recommended to study these directions in depth using the references provided here as a starting point.

References

1. Yinzhi Cao and Junfeng Yang. Towards making systems forget with machine unlearning. In *2015 IEEE Symposium on Security and Privacy*, pages 463–480. IEEE, 2015.
2. Jonathan Brophy and Daniel Lowd. Machine unlearning for random forests. In *International Conference on Machine Learning*, pages 1092–1104. PMLR, 2021.
3. Lucas Bourtoule, Varun Chandrasekaran, Christopher A Choquette-Choo, Hengrui Jia, Adelin Travers, Baiwu Zhang, David Lie, and Nicolas Papernot. Machine unlearning. In *2021 IEEE Symposium on Security and Privacy (SP)*, pages 141–159. IEEE, 2021.
4. Haonan Yan, Xiaoguang Li, Ziyao Guo, Hui Li, Fenghua Li, and Xiaodong Lin. ARCANE: an efficient architecture for exact machine unlearning. In Luc De Raedt, editor, *Proceedings of the Thirty-First International Joint Conference on Artificial Intelligence, IJCAI 2022, Vienna, Austria, 23-29 July 2022*, pages 4006–4013. ijcai.org, 2022.
5. Kevin Kuo, Amrith Setlur, Kartik Srinivas, Aditi Raghunathan, and Virginia Smith. Exact unlearning of finetuning data via model merging at scale. *arXiv preprint arXiv:2504.04626*, 2025.
6. Yonatan Dukler, Benjamin Bowman, Alessandro Achille, Aditya Golatkar, Ashwin Swaminathan, and Stefano Soatto. Safe: Machine unlearning with shard graphs. In *Proceedings of the IEEE/CVF international conference on computer vision*, pages 17108–17118, 2023.
7. Michael Kearns. Efficient noise-tolerant learning from statistical queries. *Journal of the ACM (JACM)*, 45(6):983–1006, 1998.
8. Cheng-Tao Chu, Sang Kim, Yi-An Lin, YuanYuan Yu, Gary Bradski, Kunle Olukotun, and Andrew Ng. Map-reduce for machine learning on multicore. *Advances in neural information processing systems*, 19, 2006.
9. Chong Chen, Fei Sun, Min Zhang, and Bolin Ding. Recommendation unlearning. In *Proceedings of the ACM Web Conference 2022*, pages 2768–2777, 2022.
10. Somnath Basu Roy Chowdhury, Krzysztof Marcin Choromanski, Arijit Sehanobish, Kumar Avinava Dubey, and Snigdha Chaturvedi. Towards scalable exact machine unlearning using parameter-efficient fine-tuning. In *The Thirteenth International Conference on Learning Representations*, 2025.
11. Edward J Hu, Yelong Shen, Phillip Wallis, Zeyuan Allen-Zhu, Yuanzhi Li, Shean Wang, Lu Wang, Weizhu Chen, et al. Lora: Low-rank adaptation of large language models. *ICLR*, 1(2):3, 2022.
12. Jacob Devlin, Ming-Wei Chang, Kenton Lee, and Kristina Toutanova. Bert: Pre-training of deep bidirectional transformers for language understanding. In *Proceedings of the 2019 conference of the North American chapter of the association for computational linguistics: human language technologies, volume 1 (long and short papers)*, pages 4171–4186, 2019.
13. Hugo Touvron, Thibaut Lavril, Gautier Izacard, Xavier Martinet, Marie-Anne Lachaux, Timothée Lacroix, Baptiste Rozière, Naman Goyal, Eric Hambro, Faisal Azhar, et al. Llama: Open and efficient foundation language models. *arXiv preprint arXiv:2302.13971*, 2023.

Chapter 3
Approximate Machine Unlearning

In this chapter, we discuss *approximate machine unlearning*, a practical and widely adopted family of unlearning techniques that is particularly suitable for modern large-scale models where exact retraining is often computationally prohibitive. Unlike exact unlearning methods introduced in Chap. 2, approximate unlearning aims to efficiently reduce or remove the influence of specified training samples from a trained model while preserving predictive utility of the retained data.

We first introduce the basic problem formulation and key goals of approximate unlearning, including efficiency, forgetting effectiveness, and utility preservation. Approximate unlearning methods mainly differ in how they modify the model and how they balance forgetting with maintaining performance.

We then present representative approximate unlearning methods in three main categories: gradient-based optimization approaches, model-editing based interventions, and representation-based (feature/embedding) unlearning techniques.

3.1 What Is Approximate Machine Unlearning

Approximate machine unlearning aims to efficiently solve the unlearning problem solely based on the trained model and erasing data. In contrast to exact unlearning that re-runs the learning algorithm on the retained dataset from scratch, approximate machine unlearning seeks a cheaper transformation of the already-trained model.

Given a model $\theta^\star(D)$ trained on a dataset $D = D_r \cup D_f$, where D_f denotes the forget (or erasing) set and D_r the retain set, an approximate unlearning algorithm produces updated parameters θ_u by modifying $\theta^\star$ directly, typically using only D_f (and sometimes a small amount of auxiliary or retained data). The goal is that θ_u behaves as if it had never seen D_f, while largely preserving the utility of $\theta^\star$ on D_r, but without incurring the full cost of retraining.

W. Wang, S. Yu, *Machine Unlearning: Concepts and Implementations*,
Digital Privacy and Security, https://doi.org/10.1007/978-981-92-1142-5_3

Existing approximate machine unlearning solutions can be divided into several broad families:

- **Gradient-based optimization methods.** These approaches [1–3] cast unlearning as a post-training optimization problem: starting from the trained parameters, the model is updated to attenuate the influence of the forget set D_f while maintaining performance on the retain set D_r.

  A common design principle is to make the model perform worse on the forgotten set D_f, which constraining the update so performance does not collapse on the retained set D_r. Under standard assumptions (e.g., convexity and smoothness), variants of stochastic gradient descent (SGD) can provide certified approximate unlearning guarantees, with explicit bounds on how far the resulting model may deviate from one trained from scratch.
- **Model-editing and sparsification methods.** A second line of work [4, 5] focuses on directly editing the parameter space. Recent "remain-geometry"-aware updates can embed the optimization in a Riemannian manifold induced by the retained data. The key idea is to localize and adjust parameters that carry most of the information about D_f, rather than perturbing the entire model.
- **Representation- and distillation-based methods.** A third family [6–8] targets internal representations or logits instead of raw parameters. Examples include adversarial representation erasure, where a domain classifier is trained to distinguish representations of D_f from those of retained data, and is then adversarially fooled to make the forget representations indistinguishable.

  Decision-boundary-shifting methods relocate class boundaries to push out D_f while preserving performance on D_r.

  Some contrastive-learning-based unlearning methods reshape the representation geometry around D_f.

Across these methods, approximate unlearning follows a largely shared workflow. Starting from a trained model $\theta^\star$ and a specified forget set D_f, one typically (i) optionally estimates how sensitive the model is to D_f, for example using gradients or influence scores [3, 5]; (ii) constructs an update that weakens the model's reliance on D_f while preserving performance on the retained data D_r; (iii) applies one or more updates to obtain the unlearned model θ_u; and (iv) evaluates the result using empirical audits, such as membership inference, representation similarity to retraining, or worst-case bounds derived from certified analyses [1, 2].

Overall, approximate unlearning can be viewed as a constrained model-editing problem: the goal is to efficiently move the model toward the solution one would obtain by retraining on D_r, without incurring the full cost of retraining.

3.2 Approximate Unlearning Implementation

3.2.1 Gradient-based Optimization Methods

Certified Data Removal via Newton Updates Guo et al. [3] study certified data removal for regularized empirical risk minimization under standard assumptions (strong convexity and smoothness). Their key idea is to make training slightly random in a controlled way, and then show that a single second-order update can transform the trained model into one that is (in distribution) close to the model that would have been obtained by retraining from scratch without the deleted samples. Concretely, the approach has two stages: (i) perturbed training, and (ii) Newton-style unlearning via an influence function.

Stage 1: Loss-perturbation Training (Injecting Controlled Randomness) They consider a regularized Empirical Risk Minimization (ERM) objective and add a random linear perturbation:

$$\tilde{L}(\theta; D) = \sum_{i=1}^{n} \ell(\theta^\top x_i, y_i) + \frac{\lambda n}{2}\|\theta\|_2^2 + b^\top \theta, \tag{3.1}$$

where $D = \{(x_i, y_i)\}_{i=1}^n$ denotes the training set, $x_i \in \mathbb{R}^d$ is the feature vector, y_i is the corresponding label, and $\theta \in \mathbb{R}^d$ is the model parameter. The function $\ell(\theta^\top x_i, y_i)$ is a sample-wise loss (e.g., logistic or squared loss), $\frac{\lambda n}{2}\|\theta\|_2^2$ is an ℓ_2 regularizer with strength $\lambda > 0$, included to control model complexity and to induce strong convexity. The final term $b^\top \theta$ is a random linear perturbation, where $b \in \mathbb{R}^d$ is drawn from a known distribution such as an isotropic Gaussian.

Training models with Eq. (3.1) returns the minimizer $\theta^\star = \arg\min_\theta \tilde{L}(\theta; D)$. Intuitively, the linear term does *not* change the learning problem qualitatively (the objective remains strongly convex), but it makes the output model a randomized function of the dataset. This randomized output distribution is the cornerstone of their certification argument: after deletion, the unlearning procedure must produce outputs that are statistically close to those from fresh training on the modified dataset under the same perturbation mechanism.

Stage 2: Hessian-based Removal (A One-step Newton Correction) Suppose we delete one training example $z = (x_n, y_n)$. Retraining would solve $\arg\min_\theta \tilde{L}(\theta; D \setminus \{z\})$, but doing so from scratch is expensive. Instead, Guo et al. compute a *single* Newton-like update from $\theta^\star$ that approximates the effect of removing z:

$$\theta^- = \theta^\star + H^{-1} g, \tag{3.2}$$

where g is the gradient contribution that needs to be "undone" (the regularizer term plus the deleted point's loss gradient evaluated at $\theta^\star$), and H is the Hessian matrix of the remaining-data regularized objective (also evaluated at $\theta^\star$). This is precisely the classical influence-function direction: it estimates how the ERM minimizer shifts when one example is removed, using local second-order information. Operationally, the method turns unlearning into a one-shot correction step, rather than a new training run.

Why One Step Is Enough: Bounding the Optimization Error A Newton correction is only an approximation, so the paper quantifies how close θ^- is to the true minimizer on $D \setminus \{z\}$. Under smoothness/Lipschitz conditions, they bound the first-order optimality residual on the remaining dataset, i.e., the gradient norm of the (unperturbed) regularized loss at θ^-. A small residual implies that θ^- lies near the optimum of the post-deletion objective, providing a principled notion of approximation quality.

Certification via Distributional Closeness Crucially, certification is not only about being near the new optimum. It is about being indistinguishable from retraining. The earlier loss perturbation makes the retrained model output a well-defined distribution f_A (train on $D \setminus \{z\}$ with fresh perturbation), while the unlearning mechanism induces a distribution $f_{\tilde{A}}$ (start from the original perturbation and apply the Newton correction).

By combining a bound on the Newton residual with smoothness properties of the perturbation density, one can bound the likelihood ratio between the two output distributions. In particular, there exists an event $\mathcal{G}$ with probability at least $1 - \delta$ such that, for all $\theta \in \mathcal{G}$,

$$e^{-\varepsilon} \leq \frac{f_{\tilde{A}}(\theta)}{f_A(\theta)} \leq e^{\varepsilon}. \tag{3.3}$$

Equation (3.3) yields an (ε, δ)-certified removal guarantee: the unlearned model is statistically close to a from-scratch retrained model on the dataset with the point removed.

Extensions and Practical Notes The same influence-based update extends to multiple deletions by accumulating the removed points' gradient contributions and updating the curvature accordingly. The analysis shows how the residual (and thus the certificate) degrades with the number of removals. In practice, the main computational bottleneck is applying H^{-1} to a vector. Implementations typically compute this via numerical linear algebra (e.g., solving a linear system) rather than forming H^{-1} explicitly.

Certified Machine Unlearning via Projected Noisy SGD Chien et al. [1] study certified approximate unlearning through a deliberately simple principle: use the same training primitive for both learning and unlearning. Specifically, they model training as running **P**rojected **N**oisy SGD (PNSGD) on a dataset D, which induces a distribution over model parameters. Given a deletion request that changes D to an adjacent dataset D', unlearning is implemented by running only a short additional PNSGD trajectory on D', starting from the released model. The core guarantee is distributional: after enough unlearning steps, the output distribution becomes close (in Rényi divergence) to the distribution obtained by fresh retraining on D'.

Projected Noisy SGD as the Common Primitive Let θ denote model parameters and let $f_D(\theta) = \frac{1}{n}\sum_{i=1}^{n} f(\theta; d_i)$ be the empirical objective on dataset $D = \{d_i\}_{i=1}^{n}$, optimized over a convex constraint set $C_R = \{\theta : \|\theta\| \leq R\}$. Each PNSGD step takes a (mini-batch) stochastic gradient step, adds Gaussian noise, and projects back to C_R:

$$\theta_{t+1} = \Pi_{C_R}\Big(\theta_t - \eta\, g(\theta_t; B_t, D) + \sqrt{2\eta}\,\sigma\, \xi_t\Big), \qquad \xi_t \sim \mathcal{N}(0, I). \tag{3.4}$$

Here, $\eta > 0$ is the step size, $\sigma \geq 0$ controls the noise magnitude, ξ_t is standard Gaussian noise in $\mathbb{R}^d$, and Π_{C_R} denotes Euclidean projection onto C_R.

Intuitively, the noise ensures stability at the distribution level, while projection keeps the iterates bounded. Under standard conditions for convex learning (e.g., strong convexity and smoothness), this noisy, contractive iteration mixes toward a dataset-dependent stationary distribution, so "training" can be viewed as sampling θ from that distribution after sufficient iterations.

Unlearning as a Short Continuation Run A deletion request replaces D with an adjacent dataset D', differing in one record. Rather than retraining from scratch on D', the unlearning algorithm initializes at the released model $\theta^\star$, trained on D and simply continues running the same PNSGD updates on D' for K additional steps, often using the same mini-batch schedule. The returned parameter θ^- is thus produced by the same optimizer used in learning, no Hessians or influence functions.

Certification: Distributional Closeness to Retraining The technical contribution is an analysis that connects (i) how much the PNSGD stationary distributions change when moving from D to D', and (ii) how fast the noisy projected dynamics forget their initialization. Chien et al. [1] bound the initial mismatch between the learning distributions on D and D' using a Wasserstein-style argument that captures the sensitivity of stochastic gradients, improving with larger batch size and smaller step size. They then leverage contractivity of the noisy projected iteration to show that, as the unlearning run length K increases, the divergence between (a) the unlearning output distribution (start from $\theta^\star$ and run on D') and (b) the retraining output distribution (run from scratch on D') decays geometrically. This yields an (α, ε)-Rényi unlearning guarantee, where ε scales with the data-change sensitivity and decreases as K grows, making certification possible without second-order information.

3.2.2 Model-Editing Methods

Sparsity-aided Unlearning via Prune–then–Unlearn Jia et al. [4] revisit several widely used approximate machine unlearning baselines, fine-tuning (FT) [9], gradient ascent (GA) [10, 11], Fisher forgetting (FF) [9], and influence-based unlearning (IU) [3, 12]. They asked a simple but consequential question: *does model sparsity* *make these approximate updates behave more like gold-standard retraining?* Their answer is largely affirmative: sparsity can act as a structural constraint that localizes where deletion-induced changes can accumulate, thereby reducing the discrepancy between approximate unlearning and retraining.

They extend the SGD-unrolling view of GA-style unlearning to masked (sparse) training by introducing a binary pruning mask $m \in \{0, 1\}^d$ over parameters. In this setting, unlearning updates are restricted to the active subnetwork $m \odot \theta$, where $\odot$ denotes element-wise multiplication. The deviation from the retrained solution can be bounded by (i) the displacement of the active parameters and (ii) the curvature of the loss restricted to those active coordinates. The key implication is that as sparsity increases (hence fewer degrees of freedom remain) the effective space in which unlearning can "drift" shrinks, making simple gradient-based unlearning track retraining more closely than in dense models.

Motivated by the above, Jia et al. [4] propose a prune–then–unlearn paradigm (see Fig. 3.1): *(i) prune* a trained dense model, then *(ii) unlearn* only on the resulting sparse subnetwork. Concretely, starting from a dense model θ_o trained on D, they obtain a mask m using off-the-shelf pruning procedures chosen to

(a) depend minimally on the forget set D_f,
(b) preserve predictive performance,
(c) be computationally lightweight.

In practice, they emphasize SynFlow (data-agnostic pruning at initialization) and one-shot magnitude pruning (OMP) applied to θ_o, and avoid iterative magnitude pruning (IMP) because it is more tightly coupled to the training data and substantially more expensive. After pruning, approximate machine unlearning is executed

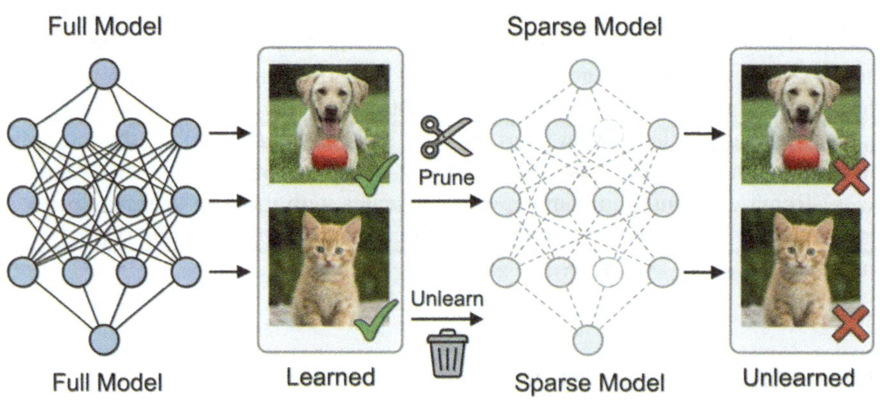

Fig. 3.1 Prune–then–Unlearn pipeline [4]. A dense model trained on D is first pruned to a sparse subnetwork (mask m) using a lightweight pruning rule, and unlearning is then performed within the masked subnetwork to remove the influence of the forget set D_f. Restricting updates to a sparse parameter subspace can reduce the gap between approximate unlearning and retraining while improving runtime

by running FT/GA/FF/IU with the mask enforced throughout (i.e., all updates are restricted to $m \odot \theta$). Empirically, this prune–then–unlearn pipeline substantially narrows the gap between approximate unlearning and retraining across CIFAR-10/100, SVHN, and ImageNet, improving forgetting/attack resistance (e.g., lower MIA success on D_f) while maintaining retain/test utility (e.g., retain accuracy and test accuracy) and reducing runtime.

Beyond the two-stage approach, Jia et al. also integrate sparsity directly into the unlearning objective by adding an ℓ_1 penalty:

$$\theta_u = \arg\min_{\theta} L_u(\theta; \theta_o, D_r) + \gamma \|\theta\|_1, \tag{3.5}$$

where L_u instantiates an unlearning loss (e.g., fine-tuning on D_r from θ_o) and $\gamma > 0$ controls sparsity.

Operationally, this corresponds to standard gradient-based updates augmented with a sparsity-inducing shrinkage effect, encouraging small-magnitude weights to move toward zero while optimizing utility on D_r. To reduce sensitivity to γ, they study simple sparsity schedules, such as constant, increasing, decreasing.

They report that a linearly decaying schedule often works best: early iterations emphasize sparsification (reducing unlearning error), while later iterations relax the penalty to refine performance on D_r. Overall, the resulting sparse machine unlearning objective achieves forgetting and privacy behavior closer to retraining, at substantially lower cost.

Saliency-Guided Model Editing via Gradient-based Weight Masks Fan et al. [5] generalize the sparsity view by introducing weight saliency as a flexible, gradient-driven way to localize where unlearning should occur in the parameter space. Instead of committing to a globally sparse subnetwork, they construct a gradient-based weight saliency map from the forgetting loss. Concretely, given a pre-trained model θ_o and forgetting set D_f, they define a forgetting loss $L_f(\theta; D_f)$ and compute its gradient with respect to all parameters, $g = \nabla_\theta L_f(\theta_o; D_f)$. Taking elementwise absolute values and applying a hard threshold (e.g., the median of $|g|$) yields a binary saliency mask s, where $s_i = 1$ marks parameters whose gradient magnitude is large and thus deemed "salient" for forgetting. The model is then conceptually decomposed into salient and non-salient components, and subsequent machine unlearning only updates the salient part $s \odot \theta$, keeping $(1 - s) \odot \theta_o$ fixed.

They instantiate the forgetting loss using gradient ascent-style objectives: for classification, L_f is the cross-entropy loss configured for gradient ascent on D_f; for diffusion models, it is based on the MSE training loss of the denoising network under the forgetting prompts. This choice makes the saliency map directly reflect "how much a parameter contributes to increasing loss on the data or concepts that should be forgotten." The resulting saliency unlearning (SalUn) framework then combines this weight saliency with random labeling (RL) to form a concrete gradient-based machine unlearning procedure.

In image classification, SalUn re-labels the forgetting examples with random incorrect labels and fine-tunes the model, but crucially restricts gradient updates to the salient weights selected by s. A regularization term penalizes deviations of the non-salient weights from their pre-unlearning values, so that generalization on the remaining data D_r is preserved while the salient sub-network absorbs most of the unlearning updates.

For image generation, SalUn adapts the same principle to conditional diffusion models. Forgetting is formulated as re-associating the harmful or undesired concept (e.g., a nudity-related prompt) with misaligned images that do not belong to that concept, while again only updating salient weights. To avoid damaging generation quality for benign prompts, the objective includes the standard diffusion MSE loss on data associated with non-forgetting prompts as a regularizer. Optimization proceeds via SGD over this combined loss, and the learned weight edits are tightly localized by the saliency mask. This allows SalUn to prevent the model from generating images for specific classes or harmful concepts, while closely matching retraining in Fréchet Inception Distance (FID) and unlearning accuracy.

Finally, Fan et al. [5] explore a soft-threshold variant, "SalUn-soft", which replaces hard masking with an ℓ_1-style saliency penalty and solves the resulting objective via proximal gradient descent, i.e., standard gradient steps followed by a closed-form soft-threshold operator on the weights. This again highlights the gradient-based optimization flavour: unlearning is driven by gradients of a carefully constructed forgetting preservation objective, and sparsity or saliency is enforced by proximal operators rather than architectural changes. In practice, the hard-thresholding SalUn typically offers a better trade-off between unlearning efficacy, stability, and preserved utility, whereas SalUn-soft demonstrates that the approach can be smoothly embedded into more general proximal optimization frameworks.

3.2.3 Representation Forgetting and Distillation-based Unlearning

Representation Forgetting with Parameter Self-Sharing (RFU-SS) Wang et al. (2024) [7] recast machine unlearning as an explicit *bi-objective* optimization problem: (i) *forgetting*, which removes the influence of erased samples, and (ii) *remembering*, which preserves utility on the remaining data. Rather than optimizing a single deletion loss and hoping utility is retained via heuristics, they proposed an algorithm that jointly controls both forgetting and remembering signals during unlearning.

Forgetting as Representation Scrubbing (RFU) The method targets forgetting at the representation level. The base model follows an information-bottleneck (IB) style decomposition $M(\theta_r, \theta_a)$: a representer θ_r maps inputs X to a latent representation Z, and an approximator θ_a predicts labels Y from Z. After training on the full dataset, unlearning is executed by driving the erased subset D_e to become uninformative with respect to the learned latent representation space. Concretely, RFU reduces the dependence between erased samples and the representations by

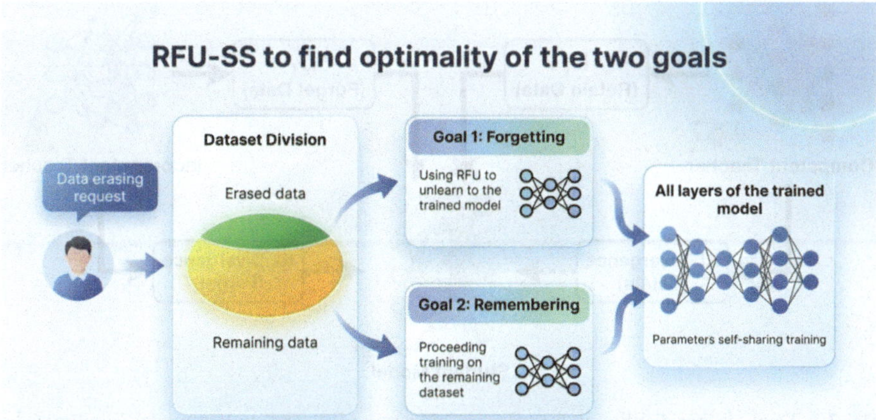

Fig. 3.2 RFU-SS overview. A deletion request partitions the training set into erased data D_e and remaining data D_r. RFU computes a representation-forgetting loss on D_e to suppress information about erased samples in the latent space, while a standard supervised loss on D_r enforces remembering. Both objectives update the *same* model parameters (parameter self-sharing), yielding a controlled trade-off between forgetting and utility preservation

minimizing mutual-information-inspired terms involving (X_e, Z_e) and (Z_e, Y_e). Since mutual information is intractable in deep models, these objectives are implemented through variational surrogates, yielding practical losses that directly discourage Z from encoding features predictive of the erased data.

Remembering via Parameter Self-sharing (SS) To preserve utility, RFU-SS couples representation forgetting with standard supervised learning on the remaining data D_r. The key design choice is parameter self-sharing: the same parameters (θ_r, θ_a) are used to optimize both the forgetting loss (computed on batches from D_e) and the remembering loss (e.g., cross-entropy on batches from D_r). This resembles multi-task learning, but avoids task-specific heads or separate subnetworks; instead, the model is steered by two gradient signals acting on all layers of the same network, as shown in Fig. 3.2.

In each iteration, RFU-SS samples minibatches from D_e and D_r, computes the forgetting and remembering gradients, and aggregates them into a single update. To prevent overly aggressive forgetting updates from destabilizing performance on D_r, the procedure incorporates simple stabilization heuristics (e.g., gradient clipping) so that the combined step does not over-dominate the remembering signal. Empirically, this joint training produces selective movement of the learned representation: features associated with D_e are pushed toward information-poor regions of Z, while the representation needed for D_r remains anchored, reducing collateral damage compared with unlearning objectives that focus solely on maximizing forgetting.

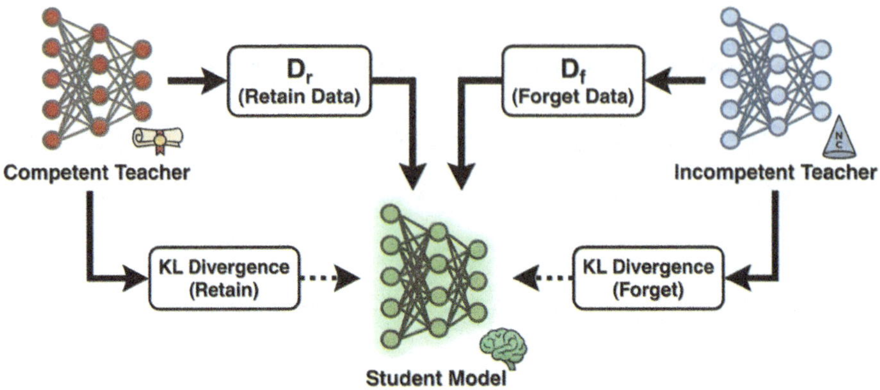

Fig. 3.3 Dual-teacher distillation for selective unlearning [6]. The student model is optimized to match a competent teacher on retained data D_r and an incompetent teacher on forgotten data D_f using KL-divergence losses, encouraging utility preservation on D_r while suppressing meaningful predictive structure on D_f

Incompetent-Teacher Representation Unlearning Chundawat et al. [6] propose a dual-teacher distillation framework in which a student model is guided to selectively forget a user-specified subset while preserving utility on the retained data. The method uses two teachers with intentionally different roles: a competent teacher that represents the original trained model (capturing well-generalized knowledge over the full training corpus), and an incompetent teacher whose predictions on the forget set are deliberately uninformative (e.g., random or weakly trained). The student is initialized from the competent teacher's parameters. It starts from the fully trained solution, and is then updated so that its outputs remain aligned with the competent teacher on retain samples D_r while being driven toward the incompetent teacher on forget samples D_f. This turns unlearning into a controlled representation-level edit implemented through distillation, without requiring full retraining, as shown in Fig. 3.3.

Distillation Objective Let the competent teacher be $T_s(\cdot;\theta_s)$ (feed input x into the teacher model T_s with parameter θ_s), the incompetent teacher be $T_d(\cdot;\theta_d)$, and the student be $S(\cdot;\theta)$. Each sample x is assigned an unlearning indicator $l_u \in \{0, 1\}$, where $l_u = 0$ denotes $x \in D_r$ and $l_u = 1$ denotes $x \in D_f$. The student is trained with a KL-based routing loss:

$$\mathcal{L}_u(x, l_u) = (1 - l_u)\,\mathrm{KL}\big(T_s(x;\theta_s)\,\|\,S(x;\theta)\big) + l_u\,\mathrm{KL}\big(T_d(x;\theta_d)\,\|\,S(x;\theta)\big). \quad (3.6)$$

Minimizing $\mathcal{L}_u$ over a mixture of forget samples and a (typically small) subset of retained samples enforces teacher-consistent behavior on D_r, while deliberately destroying predictive alignment on D_f by matching an uninformative teacher. In effect, the student's output distribution becomes close to the competent teacher on retained data but approaches randomized or "non-committal" predictions on forgotten data, which empirically removes fine-grained information associated with D_f while preserving generalizable structure supported by D_r.

Design of the Incompetent Teacher A second key design choice is how to construct $T_d(\cdot; \theta_d)$ so that it induces effective forgetting without high computational cost. The simplest option is a randomly initialized network, which yields near-uniform or arbitrarily biased logits and pushes the student toward uninformative predictions on D_f. For stronger forgetting, especially when the goal is class-level or sub-class forgetting, the authors also consider lightweight or partially trained incompetent teachers. Such teachers can yield output distributions that better approximate those of a model retrained without the forgotten subset, improving the empirical trade-off between low accuracy on D_f and high accuracy on D_r. Importantly, the framework is architecture-agnostic and can be applied to CNNs, ViTs, LSTMs, and MLPs without constraining the original training pipeline.

Evaluation Without Retraining: Zero Retrain Forgetting (ZRF) To assess forgetting without repeatedly training gold retrained models, Chundawat et al. introduce Zero Retrain Forgetting (ZRF), which compares the unlearned student's behavior on D_f to that of the fixed incompetent teacher. Operationally, ZRF is computed via a divergence (based on Jensen–Shannon divergence) between the student and the incompetent teacher on D_f: higher ZRF indicates that the student's predictions are closer to the incompetent teacher (hence more "forgotten"), whereas lower ZRF suggests persistent structure on D_f that may reflect residual information. In practice, ZRF is interpreted alongside forget-set accuracy and privacy audits to rule out degenerate failure modes, such as being systematically wrong in a consistent way that itself reveals membership in D_f.

3.3 Summary

This chapter reviewed approximate machine unlearning as a practical alternative to exact retraining equivalent deletion, emphasizing its central tradeoff between forgetting efficacy, utility preservation on retained data, and computational efficiency.

We surveyed representative implementation techniques in three categories: gradient-based optimization approaches, model-editing methods, and representation- and distillation-based methods.

These methods highlights that modern approximate unlearning methods differ mainly in where they intervene (parameters vs. saliency-masked subspaces vs. representations/logits) and how they control the forgetting–utility trade-off, providing a foundation for the evaluation and auditing strategies discussed in subsequent chapters.

References

1. Eli Chien, Haoyu Wang, Ziang Chen, and Pan Li. Certified machine unlearning via noisy stochastic gradient descent. *Advances in Neural Information Processing Systems*, 37:38852–38887, 2024.
2. Kairan Zhao, Meghdad Kurmanji, George-Octavian Bărbulescu, Eleni Triantafillou, and Peter Triantafillou. What makes unlearning hard and what to do about it. *Advances in Neural Information Processing Systems*, 37:12293–12333, 2024.
3. Chuan Guo, Tom Goldstein, Awni Y. Hannun, and Laurens van der Maaten. Certified data removal from machine learning models. In *Proceedings of the 37th International Conference on Machine Learning, ICML 2020, 13-18 July 2020, Virtual Event*, volume 119 of *Proceedings of Machine Learning Research*, pages 3832–3842. PMLR, 2020.
4. Jinghan Jia, Jiancheng Liu, Parikshit Ram, Yuguang Yao, Gaowen Liu, Yang Liu, Pranay Sharma, and Sijia Liu. Model sparsity can simplify machine unlearning. *Advances in Neural Information Processing Systems*, 36:51584–51605, 2023.
5. Chongyu Fan, Jiancheng Liu, Yihua Zhang, Eric Wong, Dennis Wei, and Sijia Liu. Salun: Empowering machine unlearning via gradient-based weight saliency in both image classification and generation. *arXiv preprint arXiv:2310.12508*, 2023.
6. Vikram S Chundawat, Ayush K Tarun, Murari Mandal, and Mohan Kankanhalli. Can bad teaching induce forgetting? unlearning in deep networks using an incompetent teacher. In *Proceedings of the AAAI Conference on Artificial Intelligence*, volume 37, pages 7210–7217, 2023.
7. Weiqi2 Wang, Chenhan Zhang, Zhiyi Tian, and Shui Yu. Machine unlearning via representation forgetting with parameter self-sharing. *IEEE Transactions on Information Forensics and Security*, 19:1099–1111, 2024.
8. Weiqi Wang, Zhiyi Tian, Chenhan Zhang, and Shui Yu. Scu: An efficient machine unlearning scheme for deep learning enabled semantic communications. *IEEE Transactions on Information Forensics and Security*, 2024.
9. Aditya Golatkar, Alessandro Achille, and Stefano Soatto. Eternal sunshine of the spotless net: Selective forgetting in deep networks. In *Proceedings of the IEEE/CVF conference on computer vision and pattern recognition*, pages 9304–9312, 2020.
10. Seth Neel, Aaron Roth, and Saeed Sharifi-Malvajerdi. Descent-to-delete: Gradient-based methods for machine unlearning. In *Algorithmic Learning Theory*, pages 931–962. PMLR, 2021.
11. Anvith Thudi, Gabriel Deza, Varun Chandrasekaran, and Nicolas Papernot. Unrolling sgd: Understanding factors influencing machine unlearning. In *2022 IEEE 7th European Symposium on Security and Privacy (EuroS&P)*, pages 303–319. IEEE, 2022.
12. Zachary Izzo, Mary Anne Smart, Kamalika Chaudhuri, and James Zou. Approximate data deletion from machine learning models. In *International conference on artificial intelligence and statistics*, pages 2008–2016. PMLR, 2021.

Chapter 4
Auditing Machine Unlearning

Having introduced algorithms for exact and approximate machine unlearning, we now turn to the complementary question of how to check whether unlearning has in fact occurred. In practice, model providers may use proprietary pipelines and users rarely have visibility into how deletion requests are executed. In the worst case, a provider could even behave strategically, giving no reliable basis for trust.

Therefore, auditing machine unlearning seeks principled procedures to test a delivered model's behaviour is consistent with a counterfactual model that was never trained on the erased data. This perspective reframes unlearning as a verifiable privacy and compliance guarantee rather than a purely algorithmic property: an auditor must account for realistic threat models, query access levels (black-box vs. white-box), and the risk of both false accusations and undetected non-compliance.

In this chapter, we first introduce the auditing problem for machine unlearning. Then, we review existing attack- and test-based approaches for detecting residual dependence on forgotten data, and discuss how auditing metrics, protocols, and assumptions interact with the underlying unlearning mechanisms.

4.1 What Is Machine Unlearning Auditing

Machine unlearning auditing includes unlearning extent evaluation and effectiveness verification. It denotes a set of techniques that enable data owners or users to assess whether and to what extent an unlearning request has actually removed the influence of a specified forget set D_f from a trained model. Unlike unlearning algorithms, which modify the model to satisfy deletion requests, auditing focuses on evidence and accountability. It constructs tests, signals, and metrics that can distinguish a faithfully unlearned model from ones that merely imitate compliance (e.g., by patching predictions on D_f), while also quantifying residual privacy risks and utility trade-offs [3, 8].

W. Wang, S. Yu, *Machine Unlearning: Concepts and Implementations*,
Digital Privacy and Security, https://doi.org/10.1007/978-981-92-1142-5_4

Table 4.1 Taxonomy of machine unlearning auditing and verification methods

Taxonomy	Core idea	Representative works
Protocol-level verifiable unlearning	Auditing is embedded into a workflow that provides user-verifiable evidence of deletion. Verification relies on protocol artifacts rather than trusting the server's claim.	VeriFi [1]; SMS [2]
Attack-based verification	Use adversarial probes (e.g., membership inference and backdoor-style tests) as post-hoc evidence: if attack success remains high after unlearning, D_f likely still influences the model. Often simple to deploy but potentially bypassable by audit-aware providers.	MIB [3, 4]; MIA-based auditing in [5]
Model-centric (influence/model-difference) audits	Verify forgetting through signals that are harder to spoof, such as indirect effects on reaction samples (influential sample pairs) or information leakage encoded in pre/post model differences, often via reconstruction-based tests.	IndirectVerify [6]; EMU [7]; TAPE [8]
Benchmark- and metric-suite evaluation	Treat auditing as systematic evaluation across diverse scenarios and metrics, such as utility, privacy, memorization, scalability, and sequential stability, usually including critiques of existing metrics and uncertainty-aware alternatives.	MUBox [9]; MUSE [10]; SeUL [11]; metric critique [12]

As summarized in Table 4.1, existing auditing methods can be organized into four broad families that differ in what they treat as evidence and how robust they are to adaptive, audit-aware providers.

A. Protocol-level Verifiable Unlearning Protocol-level schemes embed auditing into a workflow that produces user-verifiable evidence of deletion. In federated learning, Verifiable federated unlearning (VeriFi) [1] operationalises a "right to verify" procedure for departing clients: the client first performs a marking phase by injecting marker samples that fingerprint its contribution, then performs a checking phase that monitors how the global model's behaviour on these markers evolves after server-side unlearning. Complementarily, Self-supervised Model Seeding (SMS) binds verification to user-specific secrets rather than synthetic triggers: it seeds a model during training via a self-supervised auxiliary task so that a user-specific identifier is implicitly encoded, and after an unlearning request the user verifies compliance by testing whether the seed remains decodable from genuine samples [2]. The key benefit of this category is that verification relies on protocol artifacts (markers or hidden seeds) rather than trusting the provider's claim of deletion.

B. Attack-based Verification A large class of early audits repurposes privacy and security attacks as verification tools. Membership Inference via Backdooring (MIB) equips a data owner with "marked" samples whose behaviour under a backdoored model differs from that under a clean model, enabling black-box inference of whether the owner's data influenced training with theoretical guarantees [3].

In an unlearning setting, such probes can be applied before and after a deletion request: if membership inference success remains high for D_f, the model plausibly has not forgotten. Similarly, Kurmanji et al. [5] use strong membership inference attacks as key evaluation signals for user-privacy-oriented unlearning, illustrating how attack success rates can serve as quantitative audit evidence alongside retain-utility metrics [5]. While attack-based audits are intuitive and relatively easy to deploy, they can be brittle against adaptive providers who fine-tune specifically to evade a known audit, motivating more robust designs.

C. Model-centric (Influence or Model-difference) Audits To resist trivial bypass strategies (e.g., patching outputs only on D_f), recent work proposes audits that tie unlearning to harder-to-spoof signals on related samples or on the model itself. IndirectVerify constructs trigger–reaction pairs such that the trigger's presence in training induces a predictable error on the reaction sample, while correctly unlearning the trigger restores the reaction's accuracy [6]. Because the audit monitors reaction behaviour rather than the target samples being unlearned, a dishonest provider cannot trivially pass the test by only altering predictions on D_f.

A complementary line of work treats the model difference between the pre- and post-unlearning models as an information-bearing object. EMU learns reconstruction models on simulated model differences to recover attributes of the erased data, thereby quantifying residual leakage without relying on explicit backdoors [7]. TAPE further specialises this idea by learning reconstructors over tailored posterior differences, augmented with data perturbation and influence-based partitioning, to audit how much private information about forgotten samples remains recoverable under diverse unlearning scenarios [8]. Collectively, these methods shift auditing from direct probing toward principled evidence derived from influence and model-change signatures.

D. Benchmark and Metric-suite Evaluation Beyond per-request auditing, a growing body of work treats auditing as a systematic evaluation and benchmarking problem, arguing that no single metric can capture "forgetting" across different applications. MUBox integrates a broad set of unlearning methods, scenario types, and evaluation metrics into one framework, demonstrating that methods behave inconsistently across tasks and that metric choice can strongly affect conclusions [9].

For large language models, MUSE proposes a six-way evaluation benchmark spanning memorisation, privacy leakage, utility preservation, scalability, and stability under sequential unlearning, revealing substantial gaps between current methods and deployment expectations [10]. In selective forgetting for language models, SeUL introduces targeted metrics (e.g., S-EL and S-MA) to measure forgetting of sensitive spans while preserving general generation ability [11]. Moreover, Shi et al. critique commonly used metrics such as unlearning accuracy and membership inference success, showing that they can overstate forgetting, and propose conformal-ratio metrics that better detect "fake unlearning" in an uncertainty-aware manner [12].

Taken together, the taxonomy in Table 4.1 highlights that unlearning auditing has evolved from ad hoc, accuracy-only checks toward protocol-grounded verification and model-centric evidence that is more robust to audit-aware adversaries. Nevertheless, open challenges remain in designing audits that are simultaneously (i) robust under adaptive providers, (ii) scalable to large models and continual or sequential unlearning, and (iii) interpretable enough to serve regulatory and end-user needs.

4.2 Techniques of Machine Unlearning Auditing

4.2.1 Protocol-Level Unlearning Verification Schemes

VeriFi: Verifiable Federated Unlearning Gao et al. [1] propose *VeriFi*, a protocol-level framework that turns federated unlearning from a server-side claim into a client-verifiable procedure. The key idea is to couple any underlying federated unlearning algorithm with a lightweight "right-to-verify" workflow, so that a leaving client can obtain evidence that its contribution has been removed from the global model, even under limited access and without trusting the server.

In the VeriFi framework, unlearning is treated as a staged service with an explicit timeline. As illustrated in Fig. 4.1, VeriFi partitions training into a free stage $[T_0, T_{\text{enabled}})$ and an unlearning-enabled stage $[T_{\text{enabled}}, T_{\text{total}}]$. The free stage follows standard FL training to reach a high-utility global model. After unlearning is enabled at T_{enabled}, the protocol introduces additional steps—marking, checking, and server-side unlearning—that are triggered by a client leaving request and are coordinated by explicit notify messages between the client and the server.

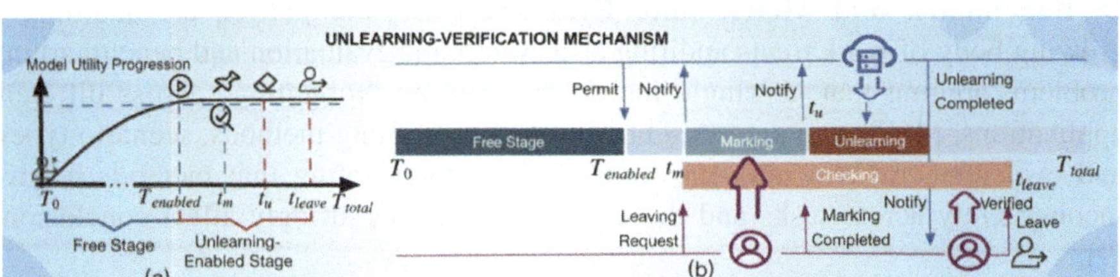

Fig. 4.1 VeriFi workflow and timeline [1]. (**a**) Training is divided into a free stage and an unlearning-enabled stage starting at T_{enabled}, with key protocol times for marking (t_m), unlearning (t_u), and client departure (t_{leave}). (**b**) After a leaving request, the client conducts marking and checking while the server performs unlearning; notifications coordinate completion and the client leaves only after verification succeeds

The core mechanism is a two-step **marking–checking** verification loop. VeriFi operationalises verification through two coupled phases executed by the leaving client: a marking phase that plants a client-specific "fingerprint" into the federated training dynamics, and a checking phase that monitors whether this fingerprint disappears after the server performs unlearning. Conceptually, the fingerprint is implemented via a small set of marker samples privately held by the client; the protocol is designed so that successful unlearning should measurably change the global model's behaviour on these markers, while a non-compliant or partially unlearned model will continue to respond to them in a distinctive way.

Marking embeds client-specific evidence into training with low overhead. During marking (around time t_m in Fig. 4.1), the leaving client performs a short, controlled interaction with the FL process that injects the marker effect into the global model through its local updates. Importantly, this step is designed to be lightweight and compatible with standard FL pipelines: it does not require modifying other clients, and it does not assume the server can access raw client data. The practical role of marking is to create a verifiable linkage between the client's contribution and a measurable signal on the eventual global model.

Checking verifies unlearning by tracking post-request marker behaviour rather than trusting server reports. After the client submits a leaving request (around T_{enabled}) and completes marking, VeriFi initiates checking: the client repeatedly evaluates the evolving global model on its marker set and tracks a verification statistic (e.g., loss/accuracy/confidence on markers) over the rounds following the unlearning request. If the server truly removes the client's contribution, the marker signal should degrade or shift in a predictable direction; if the server skips unlearning or only performs superficial output patching, the marker signal is expected to remain anomalously strong. This design anchors verification in observable behaviour of the deployed model, providing client-side evidence without requiring access to server internals.

Unlearning remains modular: VeriFi wraps around existing federated unlearning algorithms. A central design goal in [1] is separability: VeriFi is a protocol layer that can be combined with different FL unlearning implementations. In the workflow (time t_u in Fig. 4.1), the server executes an unlearning procedure on the global model parameters (denote the pre-unlearning model by θ and the post-unlearning model by θ^-), while the client independently checks whether the marker footprint vanishes as expected. This modularity allows VeriFi to benchmark multiple (unlearning, verification) combinations and to study how verification reliability depends jointly on marker design, protocol timing, and the chosen metrics.

VeriFi highlights that verification reliability depends on protocol timing and metric choice. The timeline in Fig. 4.1 makes explicit that verification is not an afterthought: the order and duration of marking/checking relative to unlearning and client departure (t_{leave}) matter for both overhead and confidence. VeriFi therefore evaluates how quickly a client can obtain sufficient evidence, how robust the signal is under different FL settings, and how the verification metric trades off false acceptance (believing unlearning happened when it did not) versus false rejection (flagging honest unlearning as non-compliant).

Within the broader auditing landscape, VeriFi exemplifies protocol-level verifiable unlearning: it provides an explicit workflow that produces client-verifiable evidence and reduces reliance on unverifiable server claims. Compared with purely attack-based audits (e.g., membership inference probes), VeriFi's emphasis is not on extracting private information, but on producing a service-level compliance test that can be executed by the leaving client under realistic access constraints.

SMS: Self-supervised Model Seeding for Verification of Machine Unlearning Wang et al. [2] propose SMS, a user-centric verification mechanism that binds a data owner's samples to a hidden "seed" during training and then checks whether the seed remains recoverable after an unlearning request. The core idea is to transform verification from an external attack (e.g., membership inference) into an intrinsic property of the trained representation: if the model has truly forgotten the user's data, it should no longer retain the user-specific seed signal that was intentionally encoded together with those samples.

Data Seeding: Embedding a Secret Without Changing the Visible Label Semantics As shown in Fig. 4.2a, the data owner constructs seeded examples by imperceptibly embedding a user-specific seed pattern S into a benign input X to obtain a seeded input X_s. This seeding can be implemented using lightweight steganographic or invisible-noise injection techniques, so that X_s is visually indistinguishable from X for humans and does not alter the primary task label. The seed is kept private by the user and is not revealed to the server, which prevents the server from trivially spoofing verification by learning an explicit detector for S.

Joint Training: A Primary Task Branch Plus a Self-supervised Seed Branch SMS then modifies training to include two coupled learning signals (the two branches in Fig. 4.2b). The primary branch learns the standard task (e.g., classification) using the seeded inputs X_s in place of (or alongside) the original inputs. In parallel, a self-supervised model seeding branch trains an auxiliary objective that encourages the network's intermediate representations to encode enough information to reconstruct or decode the seed-associated content. Concretely, the model is trained to output both the primary prediction and a seed-related representation $\hat{X}_s$ (or an embedding sufficient for seed recovery). Because both branches share the same backbone parameters θ, the seed signal becomes entangled with the representation learned from the user's training data, creating a verification handle that is "attached" to the user's contribution.

Verification Principle: Recoverability Disappears After Faithful Unlearning The verification step in Fig. 4.2c follows a simple logic: before unlearning, the trained model should enable reliable recovery/decoding of the user's seed from queries on seeded inputs. After unlearning of that user's data, the seed recoverability should collapse. Practically, the user queries the service with their seeded samples and evaluates a seed-recovery score (e.g., reconstruction fidelity or decoding success). A large drop in this score after the unlearning request is interpreted as evidence that the model no longer retains the user-specific seed binding, hence providing black-box evidence of unlearning on genuine data rather than on synthetic backdoor triggers.

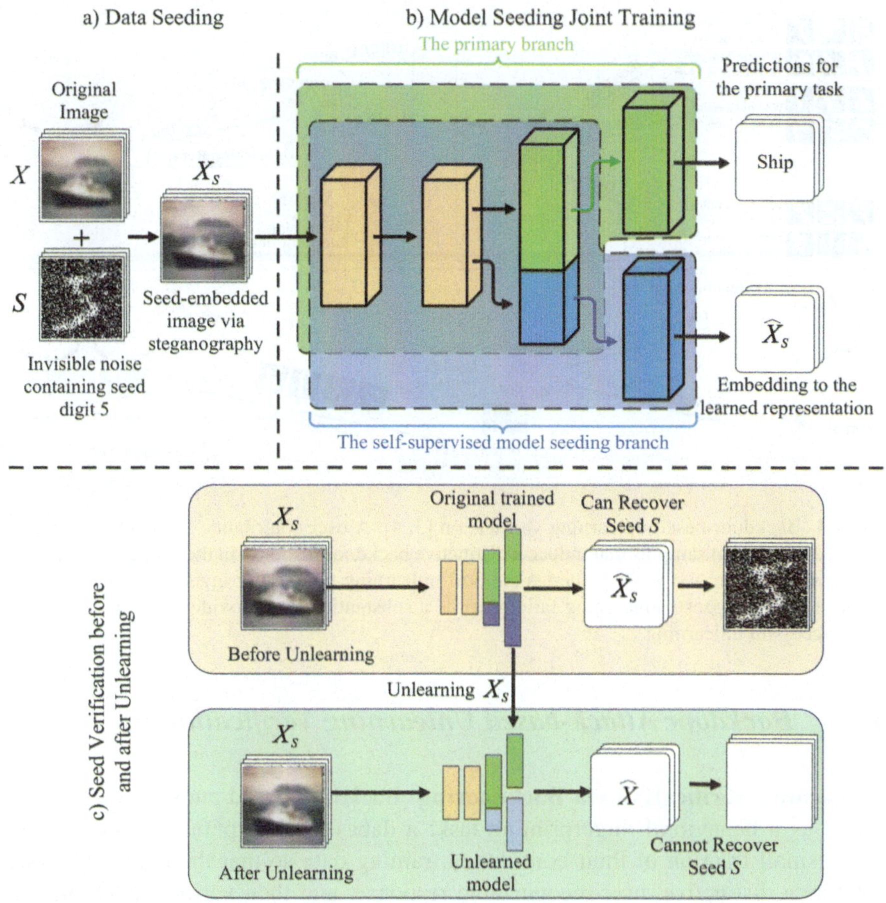

Fig. 4.2 SMS overview [2]. (**a**) A user embeds a private seed S into inputs to form seeded samples X_s. (**b**) The model is trained jointly with a primary-task branch and a self-supervised seeding branch that encourages the shared representation to encode the seed. (**c**) Verification checks seed recoverability before and after unlearning: successful unlearning makes the seed unrecoverable from the unlearned model

Why SMS Is Harder to Spoof than Direct Output Patching A key advantage of SMS is that the audit target is not simply the model's prediction on D_f (which a malicious provider could patch), but the presence of a hidden user-bound signal in the internal representation. Since the seed is private to the user and verification is performed through recovery rather than label correctness, the server cannot easily fabricate compliant behavior without either (i) actually performing effective unlearning, or (ii) knowing and modeling the user's secret seed mechanism. This shifts verification from "does the model misclassify these points now?" to "does the model still carry a user-specific representation signature that should vanish upon deletion?"

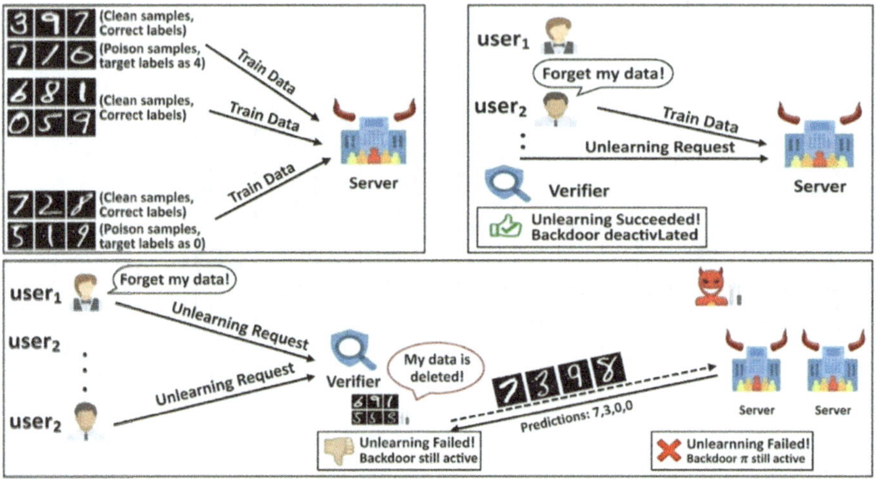

Fig. 4.3 Backdoor-based unlearning verification [3, 4]. A user contributes a small set of trigger-stamped (poisoned) samples that induce a distinctive backdoor response in the trained model. After requesting deletion, the verifier queries the post-unlearning model on triggered inputs: persistent trigger success suggests unlearning failure, while a substantial drop provides evidence consistent with successful unlearning

4.2.2 Backdoor Attack-based Unlearning Verification

Unlearning Verification via Backdooring Backdoor-based auditing casts verification as a behavioral fingerprinting task: a data owner implants a secret trigger into a small fraction of their contributed training data so that the deployed model exhibits a distinctive, user-recognizable response, and then verifies unlearning by checking whether this trigger-induced behavior disappears after a deletion request [3, 4]. As illustrated in Fig. 4.3, the verification signal is purely query-based and thus compatible with black-box access.

Verification relies on the persistence (or removal) of a trigger-induced behavior. The verifier selects a trigger pattern τ and a target output y_τ, and creates a small poisoned subset by stamping τ onto selected inputs while assigning the target label y_τ. If the poisoned subset remains influential in the learned parameters, the model will map triggered queries to y_τ with high probability; conversely, if the user's data has been faithfully removed, the trigger response should no longer be reliably reproduced. This before-and-after behavioral contrast is the basic audit signal shown in Fig. 4.3.

MIB amplifies membership evidence through backdoor controllability under black-box access. Membership Inference via Backdooring (MIB) [3] exploits the fact that backdoors can create a sharp behavioral gap between models that have versus have not incorporated a target user's contribution. In the unlearning context, the same mechanism becomes a verification test: after a "forget my data" request, a persistent backdoor response indicates that the user's data likely still affects the model, whereas the collapse of the response is consistent with successful deletion.

Operationally, unlearning verification reduces to a pre/post test on trigger success. Let θ denote the model before deletion and θ^{-} the released model after the claimed unlearning operation. The verifier evaluates the trigger behavior using an ASR-type statistic on a private trigger query set Q_τ:

$$\text{ASR}(\theta; Q_\tau) = \frac{1}{|Q_\tau|} \sum_{x \in Q_\tau} \mathbb{I}[f_\theta(x \oplus \tau) = y_\tau]. \tag{4.1}$$

Unlearning is deemed successful if $\text{ASR}(\theta^{-}; Q_\tau)$ drops substantially relative to $\text{ASR}(\theta; Q_\tau)$ (or falls below a calibrated threshold), matching the pass/fail intuition in Fig. 4.3.

Probabilistic verification formalizes the decision as hypothesis testing. Sommer et al. [4] formalize unlearning verification as a statistical decision problem between two hypotheses: H_0 (the post-unlearning model is consistent with retraining without D_f) and H_1 (residual influence of D_f remains). In a backdoor-based instantiation, the test statistic can be the post-deletion trigger response, and repeated black-box queries provide samples to estimate this statistic and report confidence rather than a purely heuristic decision.

Backdoor-based audits are attractive because they are simple, query-only, and often model-agnostic. However, an adaptive provider can potentially suppress the trigger behavior directly (e.g., via targeted fine-tuning or backdoor cleansing) without performing faithful unlearning, leading to false acceptance. This motivates stronger audit families, such as indirect trigger–reaction constructions and model-difference-based audits, that are less vulnerable to targeted evasion while retaining practical deployability.

4.2.3 *Model-Centric Auditing Methods*

In contrast to protocol-level verification and backdoor attack-based audits, model-centric auditing methods verify forgetting by analysing inevitable traces of the unlearning operation itself. The common premise is that if a subset D_f truly stops influencing the model, then this must manifest as (i) measurable downstream effects on related samples whose behavior was causally coupled to D_f, or (ii) measurable structure in the model difference between the pre- and post-unlearning models. By shifting verification away from the unlearning request samples (which an untrusted provider can often "patch" cheaply), model-centric audits aim to produce signals that are harder to spoof while remaining compatible with realistic MLaaS access constraints [6–8].

Auditing via Model Differences: EMU and TAPE Model-difference-based auditing starts from a simple but powerful premise: any unlearning request necessarily

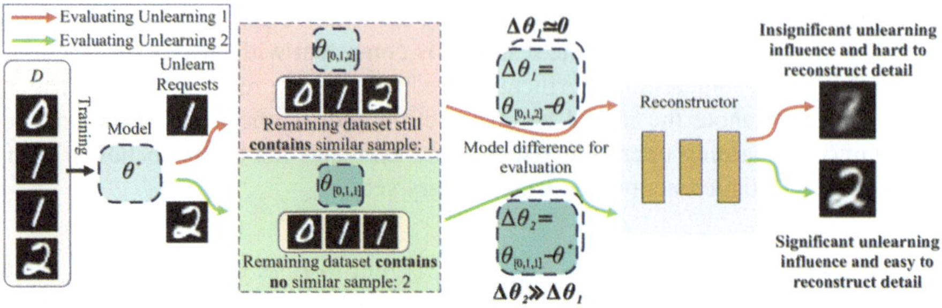

Fig. 4.4 An example of why model differences reflect data redundancy in unlearning evaluation [7]. A model trained on $D = [0, 1, 1, 2]$ receives two deletion requests: removing a redundant example (digit "1", red) versus removing a unique example (digit "2", *green*). When similar retained samples remain, the post-unlearning solution can be "explained" by the remaining data, producing a small parameter change ($\Delta\theta_1 \approx 0$) and weak reconstruction signals. When no similar sample remains, unlearning induces a much larger shift ($\Delta\theta_2 \gg \Delta\theta_1$), yielding a stronger and more reconstructable model-difference signal

produces a before-unlearned model θ and an after-unlearned model θ^-, and their discrepancy encodes how strongly the erased data shaped the learned solution [7, 8]. Rather than relying on crafted triggers, these methods treat the difference signal as the primary auditing object. A key nuance is that the magnitude and informativeness of this signal depend on redundancy in the remaining dataset: if the retained data still contains near-duplicates or highly similar samples, the optimizer can recover a similar solution without large movement, making unlearning harder to "see" through deltas. This effect is illustrated in Fig. 4.4, where removing a redundant digit produces a much smaller shift than removing a unique digit.

The Evaluation of Machine Unlearning (EMU) framework [7] operationalizes this idea by treating $\Delta\theta$ as an information-bearing artifact of training: if the erased samples contributed meaningfully to the optimization trajectory, then some of their signature should remain recoverable from the induced parameter change. EMU trains a reconstruction-based evaluator that ingests $\Delta\theta$ and attempts to recover erased attributes. Reconstruction quality is then used as a quantitative proxy for residual memorization: if the reconstructor can still extract detailed erased information from $\Delta\theta$, the unlearning procedure likely left meaningful traces; if reconstructions degrade to noise or generic patterns, the auditor interprets this as stronger forgetting [7]. The contrast in Fig. 4.4 captures the intuition behind EMU's signal: deleting a unique sample tends to induce a larger and more structured $\Delta\theta$, which is easier for a reconstructor to exploit, whereas deleting a redundant sample may yield a small delta with limited recoverable detail.

To make evaluation portable across architectures and tasks, EMU introduces a shared encoder that compresses raw model differences into a latent code optimized for downstream auditing, using an information-bottleneck-style objective to retain audit-relevant structure while discarding nuisance variations [7]. The accompanying analysis emphasizes that delta-based auditing is intrinsically sensitive to (i) similarity between erased and retained samples and (ii) machine learning task "difficulty": when the model learns only a little information from datasets, the induced difference can be small, limiting what any $\Delta\theta$-based evaluator can reliably recover [7].

TAPE [8] follows the same model-difference philosophy but targets settings where parameter access is unavailable. Instead of $\Delta\theta$, TAPE audits posterior differences—changes in predicted probability vectors between θ and θ^- on user-held queries related to the erased data. If unlearning is faithful, the model's predictions on erased-related inputs should shift in systematic ways, while predictions on unrelated retain inputs should remain stable; these structured shifts serve as the audit signal under black-box interfaces [8].

In TAPE, they also introduced a shadow unlearning and difference augmentation method. A practical barrier is generating enough paired examples of (original vs. unlearned) posteriors to train and stress-test an auditor without repeatedly executing expensive unlearning. TAPE addresses this by using influence-function approximations to synthesize many shadow unlearned models for individual samples, then computing posterior differences between the original model and these shadows to train a reconstructor [8]. To handle multi-sample requests, TAPE further introduces augmentation mechanisms to amplify weak signals and reduce entanglement across multiple erased points. TAPE evaluates unlearning via reconstruction similarity and verifiability-style scores that quantify what erased information remains recoverable from posterior shifts [8].

Takeaways EMU and TAPE demonstrate that *backdoor-free, post-hoc* auditing can be built around model differences, either in parameters or outputs [7, 8]. However, as highlighted by the redundancy effect in Fig. 4.4, difference signals are not uniformly strong: when erased samples are well-supported by similar retained data, the induced change can be small and difficult to interpret, motivating hybrid audits that combine delta signals with additional causal/protocol evidence in challenging regimes.

Verifying Machine Unlearning via Influential Sample Pairs Xu et al. [6] propose IndirectVerify, a verification scheme designed for an untrusted MLaaS threat model in which the provider may attempt to pass audits by cheap fine-tuning instead of actually unlearning. The key observation is that many existing verifiers reuse the same or closely related samples for (i) the unlearning request and (ii) the verification test; consequently, a dishonest provider can overfit to these audit points without removing the influence of D_f from the model parameters. IndirectVerify breaks this vulnerability by deliberately separating trigger samples (the data requested for deletion) from reaction samples (the data used to verify), so the provider cannot infer the audit target from the deletion request alone.

Core Idea: Influential Trigger–Reaction Coupling IndirectVerify constructs influential sample pairs (D_{trig}, x_r) such that training with the trigger set D_{trig} causes the reaction sample x_r to be misclassified, while correctly unlearning D_{trig} restores x_r's correct prediction. Conceptually, this leverages a basic property of ERM: individual training points can have non-trivial influence on the loss of specific test points, and this influence can be amplified by making a small set of training samples "harmful" to a chosen reaction point. Verification then reduces to an indirect causal check: if unlearning truly removes the trigger influence, the reaction behavior must change accordingly [6].

Pair Generation via Perturbation and Gradient Alignment To instantiate such pairs, the user perturbs a small subset of uploadable training samples to form the trigger set D_{trig} while keeping their labels unchanged. Rather than optimizing through the full training pipeline, IndirectVerify uses a practical outer-loop procedure that steers the triggers so that their training gradients align with directions that increase the loss (or induce a targeted misclassification) on the chosen reaction sample. Intuitively, the construction makes SGD updates that fit the trigger set also push the model in a direction that hurts x_r, thereby creating a reliable before-unlearning failure on x_r that can later be reversed only by removing the trigger influence [6].

Verification Protocol: Request Triggers, Query Reactions IndirectVerify operationalizes verification in two stages: (i) generation & injection, where the user uploads the dataset containing the triggers and confirms that the trained service model misclassifies the reaction sample; and (ii) unlearning & re-query, where the user requests unlearning of the triggers only and then re-queries the model on the reaction sample. If x_r returns to the correct label after unlearning, the audit concludes that the provider performed a deletion operation consistent with removing the trigger influence; if not, the audit flags likely non-compliance. Crucially, the reaction sample is never revealed as part of the unlearning request, making it difficult for the provider to pass the test by simply patching outputs on requested points [6].

4.2.4 Benchmark-Level Frameworks for Stress-Testing Unlearning Algorithms at Scale

MUBox: A Critical Evaluation Framework of Deep Machine Unlearning Li et al. [9] propose MUBox as a benchmarking and auditing framework designed to make deep unlearning methods comparable under controlled, reproducible settings. The central motivation is that prior results are often difficult to reconcile because papers mix different datasets, architectures, forgetting setups, and metrics. MUBox addresses this by providing a standardized end-to-end pipeline that (i) instantiates common unlearning scenarios, (ii) runs a diverse suite of exact and approximate unlearning methods under the same training protocol, and (iii) evaluates the resulting utility–forgetting–cost trade-offs with a fixed metric suite.

Input Module: Scenario Construction as the Unit of Control As illustrated on the left of Fig. 4.5, MUBox begins by turning a dataset into a well-defined unlearning instance via scenario-specific partitions of the training set. It implements a set of practically motivated scenarios (e.g., class-wise forgetting, forgetting from one class/all classes, worst-/best-case deletions, and depoisoning), each of which spec-

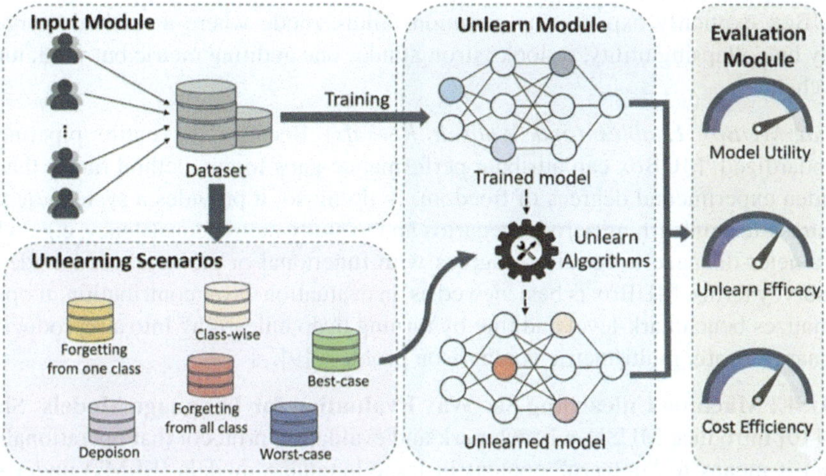

Fig. 4.5 MUBox pipeline [9]. MUBox modularizes deep unlearning evaluation into an input module (dataset and scenario construction), an unlearn module (training and unlearning algorithms producing unlearned models), and an evaluation module (utility, unlearning efficacy, and cost metrics). This design enables controlled, apples-to-apples comparisons across methods and scenarios

ifies how to form the forget set D_f and the retain set D_r. By fixing these scenario generators, MUBox ensures that different unlearning algorithms are evaluated on the same deletion task, rather than on incomparable ad hoc splits.

Unlearn Module: Producing Paired Baselines and Edited Models The middle block of Fig. 4.5 highlights the second key design choice: MUBox treats unlearning as a transformation applied to a trained base model. It first trains a reference model on the full training data, then applies a chosen unlearning algorithm to produce an unlearned model for the specified D_f. When possible, MUBox also includes retraining baselines (e.g., retrain-from-scratch, and partitioned retraining such as SISA) so that approximate methods can be compared against a gold-standard "trained only on D_r" reference under identical model and data configurations. This paired-model view (original vs. unlearned vs. retrained) is essential for studying whether a method's edits actually approximate retraining, rather than merely degrading behavior on D_f.

Evaluation Module: Multi-metric Auditing Instead of Single-number Claims On the right of Fig. 4.5, MUBox aggregates evaluation into three metric families that mirror real deployment objectives. Model utility measures how well the unlearned model retains predictive performance (e.g., on test data and on D_r). Unlearning efficacy measures whether the model's behavior on D_f aligns with the retrained reference and/or whether privacy leakage about D_f is reduced (including attack-based probes such as MIA variants, and other forget-quality signals). Cost efficiency quantifies the computational and storage overhead relative to retraining. By reporting all three, MUBox explicitly exposes the common failure mode where a method "forgets" only by collapsing utility, or looks strong under one auditing metric but weak under another.

What MUBox Enables (and What it Reveals) Because the entire pipeline is standardized, MUBox can attribute performance gaps to the method rather than to hidden experimental degrees of freedom. In doing so, it provides a systematic way to stress-test unlearning across scenarios and to study metric consistency, e.g., when parameter-distance-to-retrain disagrees with functional or privacy-based evidence. In survey terms, MUBox is best viewed as an evaluation-layer contribution: it operationalizes benchmark-level auditing by turning deep unlearning into a reproducible, scenario-aware, multi-metric comparison problem [9].

MUSE: Machine Unlearning Six-Way Evaluation for Language Models Shi et al. [10] introduce MUSE, a benchmark and evaluation protocol that operationalizes what it means to "unlearn" content in large language models (LLMs) under real deployment pressures. Rather than proposing a new unlearning algorithm, MUSE formalizes six desiderata that reflect two stakeholder views—data owners who demand removal and privacy protection, and deployers who require the model to remain useful and maintainable. As summarized in Fig. 4.6, the six criteria are: (C1) no verbatim memorization of the forget set, (C2) no knowledge memorization of its contents, (C3) no privacy leakage about whether forget examples were used for training, (C4) utility preservation on non-forgotten data, (C5) scalability as the forget set grows, and (C6) sustainability under sequential unlearning requests.

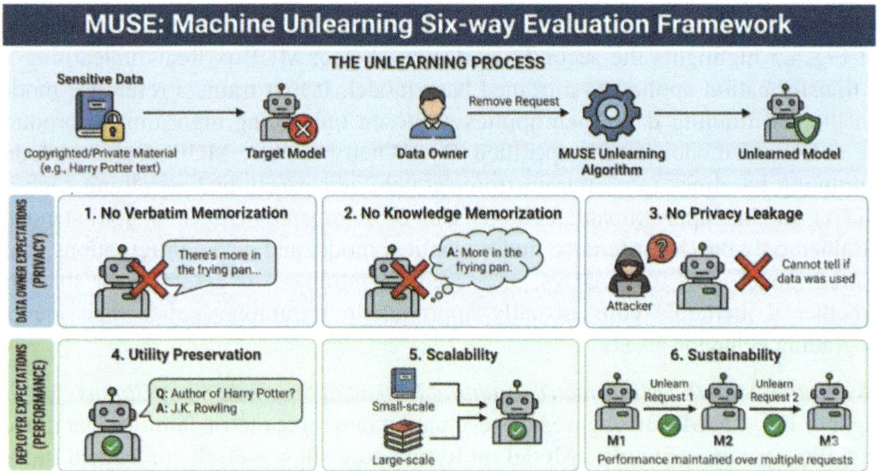

Fig. 4.6 MUSE six-way evaluation [10]. The benchmark distinguishes data-owner expectations (no verbatim memorization, no knowledge memorization, and no privacy leakage) from deployer expectations (utility preservation, scalability with respect to forget-set size, and sustainability under repeated requests), and evaluates Large Language Model unlearning methods against all six criteria

Data-owner Criteria: Separating Copying, Content Knowledge, and Privacy Leakage MUSE measures forgetting along three complementary axes. First, to test C1 (verbatim memorization), it prompts the model with a prefix from a forget-set passage and evaluates how closely the generated continuation matches the original text; high overlap indicates residual copying. Second, to test C2 (knowledge memorization), it asks questions–answer queries derived from the forgotten content and checks whether the model can still reproduce the corresponding answers, capturing non-verbatim retention of facts. Third, to test C3 (privacy leakage), it uses membership-inference-style attacks to determine whether a sample appears to have been seen during training; successful unlearning should make forget samples indistinguishable from unseen (holdout) samples under such tests.

Deployer Criteria: Utility, Scalability, and Sustainability From the deployer perspective, C4 (utility preservation) evaluates whether unlearning preserves useful capabilities on non-forgotten data, e.g., knowledge and performance on a retain set. C5 (scalability) then asks whether this utility degrades gracefully as the forget set grows from small to large removals, reflecting the reality that requests may range from individual users to large corpora. Finally, C6 (sustainability) addresses continual operation: unlearning is applied repeatedly across multiple requests, and the benchmark measures whether performance remains stable or collapses over time, mirroring production settings where deletion requests arrive sequentially.

To instantiate these criteria, MUSE constructs two representative corpora. In BOOKS, the forget set is the Harry Potter book text, while the retain set consists of Harry-Potter-related fan-wiki material that a deployer may wish to preserve as general domain knowledge. In NEWS, the corpus is built from post-cutoff news articles and split into disjoint forget/retain/holdout partitions. The benchmark derives both passage-level prompts (for verbatim memorization) and question–answer probes (for knowledge/utility), and compares unlearned models against a retrain-from-scratch baseline trained only on the retained data. This design makes the evaluation "book-like" in its logic: each criterion is tied to a concrete stakeholder concern, and each concern is tested with a targeted probe rather than a single aggregate score.

Within the common protocol, MUSE evaluates multiple families of approximate LM unlearning methods, including loss-increasing procedures on D_f (e.g., gradient-ascent-style objectives), preference-style objectives that discourage high likelihood on forgotten content, and retain-side regularizers that attempt to limit collateral damage. A recurring empirical message is that suppressing overt memorization (C1–C2) is often easier than simultaneously meeting privacy (C3) and deployer requirements (C4–C6): methods that aggressively eliminate forgetting-set signals can still leak membership cues, and they frequently incur nontrivial utility loss that worsens with larger deletions or repeated requests. In this sense, MUSE serves as a diagnostic lens: it exposes which expectations a method satisfies, which it violates, and how these trade-offs evolve as unlearning moves from toy deletions to realistic, ongoing operations [10].

4.3 Summary

This chapter surveyed machine unlearning auditing as the accountability layer that tests whether a delivered model is consistent with having never trained on the forget set D_f, often under black-box access and potentially untrusted providers.

Guided by Table 4.1, we reviewed four families of methods: (i) *protocol-level verification*, which embeds evidence into the workflow (VeriFi in Fig. 4.1, SMS in Fig. 4.2); (ii) *attack-based verification*, which uses behavioral fingerprints or membership inference but can be evaded (backdoor verification in Fig. 4.3); (iii) *model-centric audits*, which rely on indirect effects or pre/post model differences (EMU/TAPE), with Fig. 4.4 illustrating how redundancy can weaken difference signals; and (iv) *benchmark-level evaluation*, which standardizes scenarios and metrics for controlled comparison (MUBox in Fig. 4.5, MUSE in Fig. 4.6).

Overall, auditing is moving from single-metric checks toward more robust, scenario-aware evidence, while key challenges remain in robustness, scalability, and interpretability.

References

1. Xiangshan Gao, Xingjun Ma, Jingyi Wang, Youcheng Sun, Bo Li, Shouling Ji, Peng Cheng, and Jiming Chen. Verifi: Towards verifiable federated unlearning. *IEEE Transactions on Dependable and Secure Computing*, 21(6):5720–5736, 2024.
2. Weiqi Wang, Chenhan Zhang, Zhiyi Tian, and Shui Yu. Sms: Self-supervised model seeding for verification of machine unlearning. *IEEE Transactions on Dependable and Secure Computing*, 2025.
3. Hongsheng Hu, Zoran Salcic, Gillian Dobbie, Jinjun Chen, Lichao Sun, and Xuyun Zhang. Membership inference via backdooring. In Luc De Raedt, editor, *Proceedings of the Thirty-First International Joint Conference on Artificial Intelligence, IJCAI 2022, Vienna, Austria, 23-29 July 2022*, pages 3832–3838. ijcai.org, 2022.
4. David Marco Sommer, Liwei Song, Sameer Wagh, and Prateek Mittal. Towards probabilistic verification of machine unlearning. *arXiv preprint arXiv:2003.04247*, 2020.
5. Meghdad Kurmanji, Peter Triantafillou, Jamie Hayes, and Eleni Triantafillou. Towards unbounded machine unlearning. *Advances in Neural Information Processing Systems*, 36, 2024.
6. Heng Xu, Tianqing Zhu, Lefeng Zhang, and Wanlei Zhou. Really unlearned? verifying machine unlearning via influential sample pairs. *IEEE Transactions on Dependable and Secure Computing*, 2025.
7. Weiqi Wang, Chenhan Zhang, Zhiyi Tian, Shui Yu, and Zhou Su. Evaluation of machine unlearning through model difference. *IEEE Transactions on Information Forensics and Security*, 2025.
8. Weiqi Wang, Zhiyi Tian, An Liu, and Shui Yu. Tape: Tailored posterior difference for auditing of machine unlearning. In *Proceedings of the ACM on Web Conference 2025*, pages 3061–3072, 2025.
9. Xiang Li, Wenqi Wei, and Bhavani Thuraisingham. Mubox: A critical evaluation framework of deep machine unlearning. *SACMAT'25*, page 175, 2025.

10. Weijia Shi, Jaechan Lee, Yangsibo Huang, Sadhika Malladi, Jieyu Zhao, Ari Holtzman, Daogao Liu, Luke Zettlemoyer, Noah A. Smith, and Chiyuan Zhang. MUSE: Machine unlearning six-way evaluation for language models. In *The Thirteenth International Conference on Learning Representations*, 2025.
11. Lingzhi Wang, Xingshan Zeng, Jinsong Guo, Kam-Fai Wong, and Georg Gottlob. Selective forgetting: Advancing machine unlearning techniques and evaluation in language models. In *Proceedings of the AAAI Conference on Artificial Intelligence*, volume 39, pages 843–851, 2025.
12. Yingdan Shi, Sijia Liu, and Ren Wang. Rethinking evaluation metrics for machine unlearning. In *ICML 2025 Workshop on Machine Unlearning for Generative AI*.

Chapter 5
Graph Unlearning

Having introduced algorithms for exact and approximate machine unlearning, as well as auditing techniques to verify whether forgetting has truly occurred, we now turn to the specialised setting of graph unlearning.

In this chapter, we first introduce the formalization of unlearning objectives in graph-structured data, distinguish node, edge, and subgraph-level forgetting under both transductive and inductive regimes. Then, we review existing exact and approximate graph unlearning algorithms, and discuss how structural dependencies, graph dynamics, and application constraints shape the design of unlearning methods and their evaluation.

5.1 What Is Graph Unlearning

Many high-stakes applications, such as social networks [10], recommender systems [11], molecular design [12], and knowledge graphs [13], are naturally modelled as graphs, where information is encoded not only in individual nodes and edges but also in their relational structure [14]. In such settings, a single deletion request (for a node, edge, or subgraph) can propagate through message-passing and aggregation, influencing the representations of distant neighbours and global graph statistics, which makes naive data removal neither straightforward nor easily verifiable [1].

Therefore, graph unlearning asks how to adapt or redesign graph learning pipelines so that a trained model behaves as if specified nodes, edges, attributes, or motifs had never been observed, while maintaining predictive performance on the remaining graph.

Compared with independent and identically distributed (i.i.d.) data, the main complication of graph unlearning is structural dependence: because node representations are constructed by neighbourhood aggregation, changes to one part of the graph can alter the learned embeddings and predictions elsewhere [1, 4, 5]. As

W. Wang, S. Yu, *Machine Unlearning: Concepts and Implementations*,
Digital Privacy and Security, https://doi.org/10.1007/978-981-92-1142-5_5

Table 5.1 Compact taxonomy of graph unlearning methods

Taxonomy	Core idea	Representative work
Efficient retraining systems	Achieve "retrain-on-modified-graph" behaviour via partial retraining (graph sharding/partitioning, sub-model retraining, and aggregation), improving scalability while keeping behaviour close to retraining in practice.	Chen et al. [1]; Zhang et al. [2]; Xia et al.[3]
Certified approximate unlearning	Provide provable closeness to retraining on the modified graph, e.g., residual bounds and (ε, δ)-style certification, typically via analytic updates plus controlled perturbation.	Chien et al. [4]; Pan et al. [5]
Heuristic/editing-based unlearning	Fast approximate unlearning through parameter edits or influence-based updates without full retraining; typically empirically validated rather than formally certified.	Wu et al. [6]; Yang et al. [7]; Zhang et al. [8]; Wang et al. [9]

a result, unlearning methods must account for both the deletion target (node, edge, subgraph, or whole graph) and the propagation mechanism that spreads its influence.

Table 5.1 summarises a compact three-way taxonomy of graph unlearning methods. First, efficient retraining systems aim to approximate the "retrain-on-modified-graph" counterfactual through graph partitioning and partial retraining. Representative systems such as GraphEraser shard the training graph into subgraphs and retrain only affected shards upon a deletion request, followed by aggregation of sub-models [1]. Subsequent work improves sharding and aggregation to better preserve structural and semantic information during partial retraining [2]. Although the CAUSE method [3] is not graph-specific, it illustrates how exact retraining-based unlearning can be made practical under memory/energy constraints, which is directly relevant when deploying retraining-style graph unlearning at the edge.

Second, certified approximate graph unlearning prioritises formal guarantees that the post-unlearning model is close to the model obtained by retraining on the modified graph. Chien et al. introduce certified graph unlearning for node classification under simplified propagation operators, deriving analytic parameter updates with residual bounds that can be converted into (ε, δ)-style certification via perturbation mechanisms [4]. Pan et al. extend the certified removal perspective to graph classification in limited-data regimes by unlearning in a linear classifier built atop non-trainable graph scattering representations, yielding explicit residual bounds for node- and graph-level deletion requests [5]. These methods trade generality for principled, auditable guarantees.

Third, heuristic or editing-based unlearning seeks fast, practical removal without full retraining and typically evaluates success empirically rather than via certification. The GIF method adapts influence-function ideas to message passing by estimating the effect of nodes or edges through local receptive fields and performing one-shot parameter updates [6]. Complementarily, Erase-then-Rectify performs training-free parameter editing: it first erases activation patterns associated with the forget set and then rectifies the model to recover retain-set utility [7]. Related work also treats graph structure as the unlearning target, explicitly balancing forgetting and remembering by constrained pruning/rewiring of edges [8], and extends the setting to inductive deployment where the unlearned model must generalise to unseen graphs [9]. Together, these approaches occupy the fast end of the spectrum, often at the cost of weaker guarantees.

Across these strands, graph unlearning can be understood as the problem of efficiently approximating or implementing retrain-on-modified-graph behaviour under the unique constraints of graph-structured data. Certified approaches emphasise provable closeness to retraining; retraining systems provide scalable counterfactual behaviour via partial retraining; and heuristic/editing-based methods offer lightweight, deployment-friendly updates (Table 5.1). We introduce the detailed implementations of these graph unlearning methods as follows.

5.2 Graph Unlearning Implementation

5.2.1 Efficient Graph Retraining System

Graph Unlearning (GraphEraser) Figure 5.1 illustrates the key idea of GraphEraser: to make unlearning feasible on a single large training graph, the system replaces "retrain on the modified graph" with shard-local retraining plus learned aggregation of shard predictors [1]. This design adapts the SISA [15] philosophy of partition–retrain–aggregate to the non-i.i.d. dependencies of graph data, where naively partitioning nodes can destroy connectivity patterns that message-passing GNNs rely on [1, 15].

GraphEraser starts from an attributed training graph $G_0 = \langle V, A, X \rangle$, where V is the set of nodes, A denotes the adjacency information (i.e., the edge structure

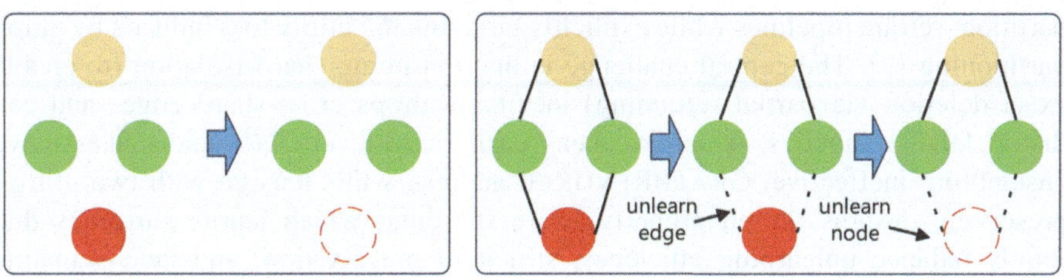

(a) Regular data unlearning (b) Graph unlearning

Fig. 5.1 Overview of GraphEraser for graph unlearning [1]. The training graph is partitioned into balanced shards; a shard-specific GNN F_i is trained per shard; predictions are aggregated by learned weights α; and an unlearning request triggers shard-local retraining (and optionally re-learning α) rather than full retraining

of the graph), and X is the node-feature matrix, and then partitions G_0 into k subgraphs (shards) $\{G_s^i\}_{i=1}^k$ (left side of Fig. 5.1) [1]. The figure contrasts two requirements that are usually in tension: (i) preserving structure so that each shard retains meaningful local neighbourhood information for GNN training, and (ii) balancing shard sizes so that the worst-case unlearning cost is bounded. To reconcile these, the authors propose balanced graph partitioning procedures that begin from structure-aware clustering or embedding-based clustering and then explicitly enforce a shard-size constraint via reassignment, yielding shards that are both cohesive and approximately equal-sized.

After sharding, GraphEraser trains one shard model $F_i(\cdot)$ per shard (middle of Fig. 5.1) [1]. For the prediction task, a query node w is evaluated by all shard models, producing k posterior vectors (dashed arrows). Rather than using uniform averaging or majority vote, GraphEraser learns a convex importance vector $\alpha = (\alpha_1, \ldots, \alpha_k)$ (right side of Fig. 5.1) so that the final prediction is a weighted aggregation $\sum_i \alpha_i F_i(w)$ [1]. The aggregation weights are fit on a small calibration subset by minimising cross-entropy under simplex constraints ($\alpha_i \geq 0$, $\sum_i \alpha_i = 1$), which makes aggregation inexpensive to update and robust to heterogeneous shard quality.

Unlearning is handled by locality: an unlearning request u (green arrow in Fig. 5.1) affects only the shard that contains the target node/edge (highlighted shard), so GraphEraser removes the requested item and retrains only the corresponding shard model F_j (bottom pathway) [1]. If the deleted item participated in learning α, the system additionally re-learns the aggregation weights; otherwise, α is reused. Because shards are balanced, each request triggers at most one shard retraining, yielding substantial speedups over full retraining while keeping utility close to the retrain-on-pruned-graph baseline across datasets and GNN backbones.

Finally, GraphEraser is explicitly an approximate graph unlearning system: it targets practical efficiency and high post-unlearning accuracy, but does not provide formal certified-removal guarantees. Instead, it evaluates residual leakage empirically using unlearning evaluation protocols.

GraphRevoker Zhang et al. propose GRAPHREVOKER, a retraining-based graph unlearning framework that retains the strong deletion guarantee of SISA-style partition–retrain pipelines while explicitly targeting the utility loss induced by graph partitioning [2]. The central challenge is that enforcing shard isolation (to enable exact deletion via partial retraining) inevitably drops cross-shard edges and can distort label semantics, which weakens each shard-level GNN and makes naive ensembling ineffective. GRAPHREVOKER addresses this tension with two utility-preserving designs: graph property-aware sharding, which learns partitions that jointly balance unlearning efficiency, structural preservation, and label-semantic richness; and graph contrastive sub-model aggregation, a lightweight contrastive ensemble module that compensates for information lost at shard boundaries. An overview is shown in Fig. 5.2.

The first major component of GraphRevoker is a graph property-aware sharding module (Fig. 5.2(1)) that partitions the training graph into subgraphs by jointly

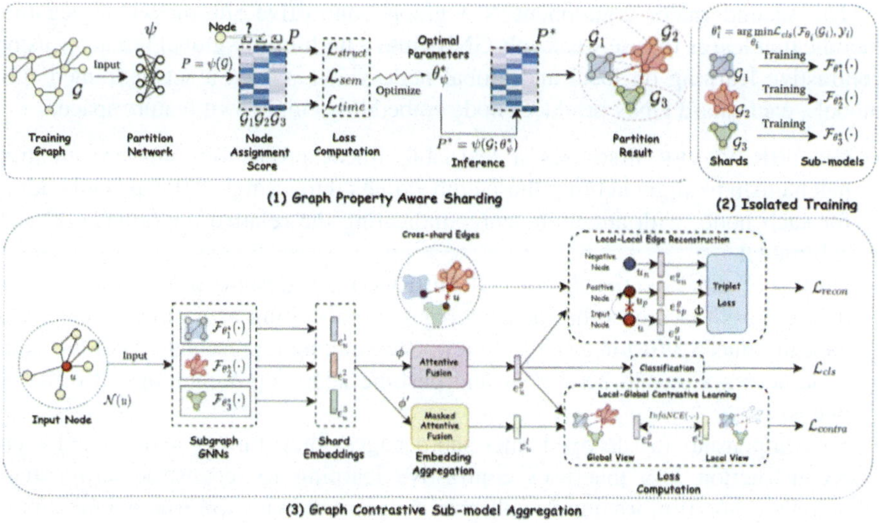

Fig. 5.2 Overview of GRAPHREVOKER [2]. (1) *Graph property-aware sharding*: a learnable partition network produces soft node–shard assignments by jointly optimising expected unlearning cost ($\mathcal{L}_{\text{time}}$), structural preservation via normalised edge-cut ($\mathcal{L}_{\text{struct}}$), and label-semantic richness via shard label-entropy ($\mathcal{L}_{\text{sem}}$), and then discretises them into shards $\{G_1, G_2, G_3\}$. (2) *Isolated training*: an independent sub-GNN is trained per shard, so a deletion request is handled by retraining only the affected sub-model(s). (3) *Graph contrastive sub-model aggregation*: projected shard embeddings are fused with attention, and the aggregator is trained with a classification objective ($\mathcal{L}_{\text{cls}}$) plus local–global contrastive learning ($\mathcal{L}_{\text{contra}}$) and local–local cross-shard neighbourhood reconstruction ($\mathcal{L}_{\text{recon}}$), recovering global utility despite cross-shard edge removals

optimising unlearning time, structural preservation, and label semantic preservation. On the efficiency side, they formalise an "unlearning time" objective that minimises the expected retraining cost over sub-models, where the probability of retraining a shard grows with its size and the per-shard cost scales with its edge count; this encourages balanced subgraphs that keep worst-case retraining time low.

On the structure side, they introduce a supervised objective based on the normalised edge-cut, which explicitly penalises edges cut between subgraphs and thus quantifies structural loss in the partitioning stage. On the semantic side, they define an entropy-based label preservation term that favours shards with rich, balanced label distributions, thereby avoiding the tendency of clustering-based partitions to group same-label nodes into a single subgraph. These three objectives are combined in a differentiable graph partition network that outputs a soft node–shard assignment matrix and is trained to minimise the composite loss. After training, the soft assignments are discretised to yield property-aware shards on which sub-GNNs are independently trained.

The second major component is a graph contrastive sub-model aggregation module that learns to fuse weak sub-GNNs into a high-utility global predictor using contrastive learning on local and global node views. Given a set of trained sub-models, each shard GNN produces node embeddings in its own feature space.

- GraphRevoker first uses a learnable linear projection plus an attention mechanism to align and fuse these embeddings into a single "global" embedding for each node, with attention scores indicating the relative importance of each sub-model.
- On top of this attentive fusion, GraphRevoker designs a local–global contrastive loss that treats the fully aggregated embedding as a global view and a partially masked fusion as a local view, encouraging representations of the same node across views to be similar and pushing apart representations of different nodes.
- To compensate for dropped inter-shard edges, they further add a local–local reconstruction loss that uses contrastive learning to reconstruct cross-shard neighbourhood information, with positive pairs drawn from true neighbours in other subgraphs and negatives from unconnected nodes.
- The final aggregation objective combines attentive fusion, local–global consistency, and local–local reconstruction.

Empirically, GraphRevoker improves the accuracy and efficiency trade-off of retraining-based graph unlearning over prior partition–retrain baselines. Across multiple benchmark graphs and common GNN backbones, it yields consistently higher predictive performance than earlier SISA-style and graph partition-based unlearning frameworks, while maintaining efficient deletion updates via partial retraining. Moreover, its retraining-time-aware sharding design can reduce unlearning latency relative to baselines, and the aggregation module typically requires only a small number of epochs to retrain after each unlearning operation, keeping the end-to-end pipeline practical at scale [2].

5.2.2 Certified Approximate Graph Unlearning

Certified Machine Unlearning on Graphs via Convex Surrogates A common route to certified unlearning on graph-structured problems is to (i) restrict the predictive model to a strongly convex objective on top of a fixed graph representation, and (ii) approximate post-deletion retraining with a single Newton-style correction whose residual error can be explicitly upper-bounded and then masked by calibrated randomness [4, 5, 16]. This design makes certification tractable because both the parameter update and its approximation gap admit closed-form control, while still capturing graph effects through the chosen representation. As illustrated in Fig. 5.3, graph deletions are fundamentally more intricate than i.i.d. sample deletions because edits to node features or topology can propagate through neighbourhood aggregation.

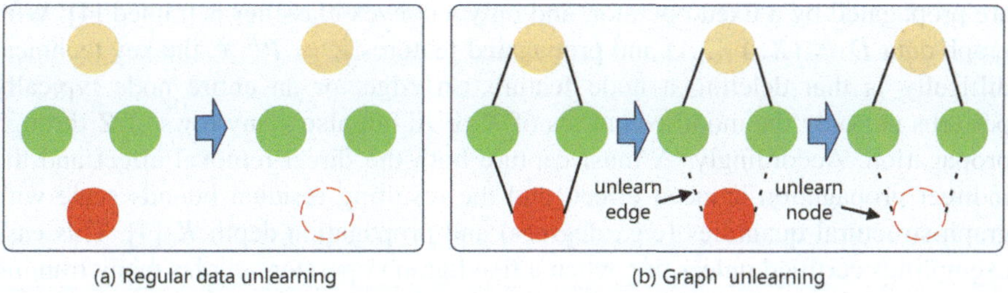

Fig. 5.3 Comparison between unlearning with i.i.d. examples and unlearning on graphs. In graphs, deletion requests may involve node features, edges, or entire nodes (with incident edges), and their effects can propagate through neighbourhood aggregation

Certified Removal Definition (Distributional Indistinguishability) Following Guo et al.'s certified removal perspective [16], the learning algorithm (including its randomness) induces a distribution over trained parameters. An unlearning mechanism is certified if, after a deletion request, the distribution of its output model is boundedly close to the distribution induced by full retraining on the post-deletion dataset [4, 5]. Conceptually, this is a DP-like stability requirement tailored to the unlearning setting: instead of requiring stability under arbitrary record changes without updates, certified unlearning allows an explicit update, but requires that the updated model remain distributionally indistinguishable from retraining.

A Unified Newton-style Certified Unlearning Template Both Chien et al. and Pan et al. instantiate certification with the same computational skeleton: start from the original minimiser $\theta^\star = \arg\min_\theta L(\theta; D)$ of a smooth, ℓ_2-regularised (hence strongly convex) objective, let D' be the dataset after deletion, and approximate the retrained optimum on D' by linearising the post-deletion optimality condition around $\theta^\star$ [4, 5]. The mechanism computes a gradient shift

$$\Delta = \nabla_\theta L(\theta^\star; D) - \nabla_\theta L(\theta^\star; D') \tag{5.1}$$

and outputs a one-step Hessian correction

$$\theta^- = \theta^\star + H^{-1}\Delta, \qquad H \approx \nabla_\theta^2 L(\theta^\star; D'). \tag{5.2}$$

Certification then hinges on bounding the approximation residual (e.g., the post-update gradient norm) $\|\nabla_\theta L(\theta^-; D')\|$, and choosing objective-perturbation noise (such as adding $b^\top\theta$ with Gaussian b) large enough to dominate this residual bound, so the deterministic discrepancy between θ^- and exact retraining is statistically hidden [4, 5, 16]. In short, "one-step update + residual bound + calibrated randomness" yields explicit (ε, δ) removal guarantees.

Instantiation I: Transductive Node-level Graph Unlearning with Propagation-induced Non-locality Chien et al. target node-level prediction on a single attributed graph under simplified graph convolution (SGC) and GPR variants, where features are propagated by a fixed operator, and only a convex classifier is trained [4]. With graph data $D = (X, Y_{\mathrm{Tr}}, A)$ and propagated features $Z = P^K X$, the key technical difficulty is that deleting a node feature, an edge, or an entire node typically perturbs not only the modified entries of X or A, but also many rows of Z through propagation. Accordingly, Δ must capture both the direct removal effect and the indirect propagation-induced effect, and the resulting residual bounds scale with graph-structural quantities (e.g., degrees) and propagation depth K [4]. This case exemplifies certified unlearning when a fixed graph operator couples many training signals, making single deletions inherently non-local.

Instantiation II: Graph-classification Unlearning with Representation Locality via GST Pan et al. consider supervised graph classification in the limited-data regime and replace the trainable GNN encoder with the Graph Scattering Transform (GST), a fixed nonlinear embedding map that produces per-graph features z_i and reduces training to regularised linear prediction on $Z = [z_1; \ldots; z_n]$ [5]. Here, the same Newton-style update applies verbatim, but the source of tractability is different: because GST is non-trainable, a deletion that modifies only one training graph changes only its corresponding row(s) in Z, leaving all other embeddings unchanged. Therefore, certification reduces to controlling how much a local edit can move the embedding, i.e., bounding $\|z_i' - z_i\|$ via GST stability properties, which directly feeds into residual bounds and noise calibration for objective perturbation [5]. This case illustrates certified unlearning by enforcing sample-level locality through a fixed representation, avoiding the global coupling induced by end-to-end trained encoders.

Takeaway Viewed together, these works show that certified graph unlearning is most tractable when learning can be expressed as a strongly convex problem on top of a fixed representation, enabling explicit one-step retraining approximations and residual-driven noise calibration [4, 5, 16]. Their main distinction is where graph dependence lives: Chien et al. embed topology through fixed propagation P^K, making deletions propagate across nodes, whereas Pan et al. enforce locality across graphs by fixing the encoder (GST), so deletions affect only a small subset of rows in the design matrix. This contrast clarifies a broader principle for certified unlearning on structured data: certification becomes easier as representation changes induced by a deletion become more local and more tightly bounded, even if the representation itself remains nonlinear and expressive.

Balanced Graph Structure Unlearning Complementary to certified unlearning methods that obtain distributional removal guarantees by restricting training to convex surrogates, Zhang et al. study the practically common setting where one must unlearn graph structure in expressive GNNs while preserving downstream utility [8]. They argue that structure unlearning is intrinsically a two-goal problem: (i) forget target-related structural influence (so that target nodes are effectively decoupled from the graph signal), yet (ii) remember performance-relevant structure for retain nodes (so that predictive accuracy does not collapse). To make this trade-off explicit, they propose SUMMIT, which formulates structure unlearning as a bi-objective optimisation rather than a single scalar loss with a manually tuned regulariser.

Bi-objective Formulation: Forgetting vs. Remembering SUMMIT optimises two losses with respect to the GNN parameters θ_{gnn}:

$$\min_{\theta_{\text{gnn}}} \left(\mathcal{L}_{\text{fgt}}(\theta_{\text{gnn}}) \,,\; \mathcal{L}_{\text{rmb}}(\theta_{\text{gnn}}) \right), \tag{5.3}$$

where the forgetting objective $\mathcal{L}_{\text{fgt}}$ encourages the removal of structural influence associated with the unlearning targets, and the remembering objective $\mathcal{L}_{\text{rmb}}$ constrains the model to preserve useful structure for retain nodes [8]. This framing aligns with the intuition that aggressive structural edits can improve forgetting but often distort neighbourhood information that the model relies on for generalisation.

Forgetting Objective with Higher-order Structure Awareness A key challenge in structural unlearning is that message passing can reintroduce dependence through multi-hop aggregation, exacerbated by oversmoothing (representations in the same connected region becoming overly similar). Therefore, Zhang et al. design $\mathcal{L}_{\text{fgt}}$ to strengthen decoupling beyond immediate neighbours by enhancing the influence of structure edits across higher-order receptive fields, so that target nodes and their neighbourhoods become more distinguishable after unlearning [8]. Intuitively, the goal is not merely to remove a few local edges, but to prevent target-related information from re-entering through multi-hop propagation.

Remembering Objective via Representation Preservation on Retain Nodes To protect utility, $\mathcal{L}_{\text{rmb}}$ penalises unnecessary drift in the latent space induced by structural modifications. Concretely, it encourages the post-unlearning GNN to maintain similar representations (in distribution) to those produced by the original well-trained model on *retain* nodes, thereby preserving the geometry that supports accurate prediction [8]. This is a knowledge-preservation constraint at the representation level: it does not freeze the adjacency or parameters directly, but restricts how much the latent embeddings of non-target nodes are allowed to change.

ATOB: Adaptive Balancing via Multi-objective Descent Rather than selecting a fixed trade-off coefficient, SUMMIT introduces an adaptive two-objective balancer (ATOB) that dynamically reconciles the two gradients during optimisation. As illustrated in Fig. 5.4, ATOB takes the per-objective gradients $g_{\text{fgt}} = \nabla_{\theta_{\text{gnn}}} \mathcal{L}_{\text{fgt}}$ and $g_{\text{rmb}} = \nabla_{\theta_{\text{gnn}}} \mathcal{L}_{\text{rmb}}$, and uses a multiple-gradient descent strategy to compute a common update direction together with an adaptive coefficient λ that balances the two objectives [8]. This mechanism steers training toward Pareto-efficient solutions, avoiding regimes where optimisation over-commits to forgetting (large utility loss) or over-commits to remembering (incomplete structural removal).

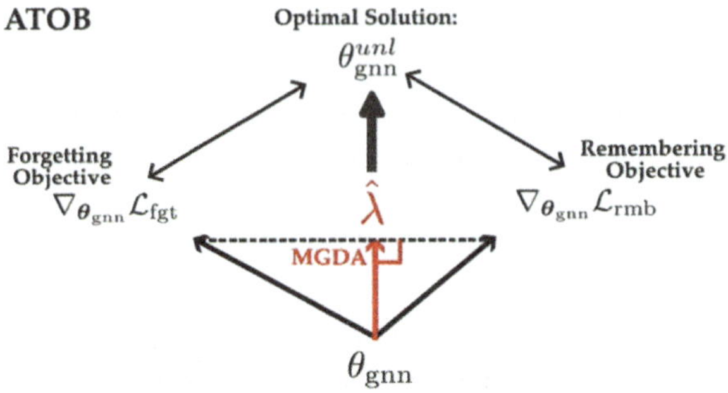

Fig. 5.4 ATOB [8] balances the forgetting and remembering objectives by adaptively combining their gradients. Given $g_{\text{fgt}} = \nabla_{\theta_{\text{gnn}}}\mathcal{L}_{\text{fgt}}$ and $g_{\text{rmb}} = \nabla_{\theta_{\text{gnn}}}\mathcal{L}_{\text{rmb}}$, ATOB uses a multiple-gradient descent algorithm (MGDA) to determine a coefficient λ and a joint descent direction that trades off the two objectives

5.2.3 Heuristic-based Graph Unlearning

A Training-Free Parameter Editing Approach for Cost-Effective Graph Unlearning Yang et al. propose Erase then Rectify (ETR), a two-stage, training-free parameter editing framework for graph unlearning that avoids iterative retraining and does not require access to the full training data after the initial model is obtained [7]. The central idea is to treat graph unlearning as selective parameter intervention: first, directly attenuate the parameters θ that most strongly encode the forgetting set (and its propagated neighbourhood influence), and then apply a lightweight one-step correction to recover utility on the remaining data. This design targets the practical bottlenecks of graph unlearning—large-scale graphs, repeated deletion requests, and non-local influence through message passing—while keeping the update cost close to a single pass of parameter editing plus a single gradient-style refinement.

Problem Setup and Parameter-importance Lens Let D denote the original training data on a graph, D_f the forgetting set (nodes/edges/features requested for removal), and D_r the remaining data after deletion. ETR's foundation is an information-theoretic view of parameter importance based on the (diagonal) Fisher information matrix (FIM): by approximating the FIM with its diagonal entries, the method obtains per-parameter sensitivity scores F_D, F_f, F_r that quantify how strongly each coordinate of θ contributes to fitting D, D_f, and D_r, respectively [7]. This decomposition makes the unlearning goal concrete: parameters with high importance for D_f but low importance for D_r are precisely those whose modification is expected to reduce memorisation of the forgetting set with minimal collateral damage to retained performance.

Erase: Neighbourhood-aware Fisher-guided Editing of θ The Erase stage edits the trained parameters θ into an intermediate vector $\hat{\theta}_r$ by selectively shrinking coordinates that are disproportionately associated with D_f (and with nodes influenced by D_f through message passing) [7]. Instead of simply zeroing all parameters with large F_f, ETR uses relative importance ratios (e.g., F_f/F_D and F_r/F_D) to separate three roles: (i) forgetting-critical parameters that primarily support D_f, (ii) propagation-affected parameters that mediate neighbour influence induced by D_f, and (iii) retain-critical parameters that mainly support D_r.

The edit rule is implemented as a piecewise scaling of coordinates of θ: forgetting-critical parameters are strongly attenuated; propagation-affected parameters are moderately adjusted, and retain-critical parameters are left nearly unchanged [7]. In practice, ETR selects these parameter sets via adaptive percentile thresholds (e.g., top-$p\%$ of the ratio scores) and introduces balancing coefficients to control how aggressively the method forgets the target samples versus their neighbourhood influence. Crucially, this step is training-free: it operates directly on θ using statistics available from the original training and from the small forgetting subset, producing $\hat{\theta}_r$ without any additional optimisation epochs.

Rectify: One-step Utility Recovery on D_r via an Approximate Gradient Because Erase is intentionally conservative on retain-critical parameters, ETR assumes that the loss landscape on D_r around $\hat{\theta}_r$ is close to that around the original optimum, and thus a single corrective update can recover much of the utility loss [7]. The Rectify stage constructs an approximation to the gradient on the remaining data, $\nabla_\theta L(\hat{\theta}_r; D_r)$, without traversing the full graph: it reuses (i) stored gradients from the original training run and (ii) correction terms computed on a small induced subgraph covering the forgetting targets and their neighbourhood. Using this approximate gradient, ETR performs one gradient-descent step,

$$\theta' = \hat{\theta}_r - \eta\, \widehat{\nabla_\theta L(\hat{\theta}_r; D_r)}, \tag{5.4}$$

yielding a rectified parameter vector θ' that better fits the retained data while preserving the forgetting effect produced by Erase [7]. Importantly, Rectify remains lightweight: it avoids full-data access, avoids multi-epoch retraining, and can be executed efficiently for repeated unlearning requests.

Complexity and Empirical Trade-offs ETR's computational profile is dominated by (diagonal) FIM estimation and coordinate-wise parameter editing, which scale linearly with the number of parameters (and, for neighbourhood corrections, with the size of the induced subgraph), and are amenable to parallelisation [7]. Across multiple node-classification benchmarks and backbones (e.g., GCN/GAT), Yang et al. report that ETR achieves a favourable unlearning–utility–efficiency balance: it attains predictive performance comparable to or better than representative baselines while delivering substantial speedups over full retraining and reducing the need to retain the full training data. Overall, ETR exemplifies a practical graph-unlearning paradigm in which structured, Fisher-guided edits to θ deliver most of the forgetting, and a single approximate corrective step restores utility with minimal additional cost.

Inductive Graph Unlearning Wang et al. propose GUIDE (**GU**ided **I**nductiv**E** Graph Unlearning), the first shard-based retraining framework explicitly designed for inductive graph unlearning, where the training procedure observes only a training subgraph and test-time nodes/edges may appear later and were never seen during training [9]. This inductive constraint breaks a core assumption behind prior transductive unlearning systems (e.g., GraphEraser), which can exploit access to the full graph (including test nodes) when partitioning, retraining, and learning aggregation rules.

GUIDE argues that naively transplanting transductive pipelines to the inductive regime leads to two failure modes: partitions become imbalanced (causing unpredictable and often high update cost) or become structurally destructive(severely harming message passing utility because many cross-shard edges are cut) [9]. To address these issues, GUIDE redesigns the unlearning pipeline around three inductive-specific objectives: (i) generate fair and balanced shards with controlled size and label distribution, (ii) preserve shard-model utility by repairing local structure lost to partitioning without leaking information across shards, and (iii) perform efficient aggregation that does not require learning global weights that would need repeated retraining under deletions.

Overall Pipeline (Partition–Repair–Update–Aggregate) As shown in Fig. 5.5, GUIDE implements inductive unlearning as a modular pipeline. It first partitions the training graph into multiple shards, then performs efficient subgraph repairing within each shard so that shard-level GNNs can still exploit meaningful neighbourhood signals after cross-shard edges are removed. Each shard trains its own GNN; upon a deletion request (node/edge/feature), GUIDE updates only the shard(s) containing the affected training elements rather than retraining globally. At inference time, GUIDE aggregates shard predictions via a similarity-based weighting rule computed between the test graph and each shard, which is lightweight to recompute after shard updates and avoids the recurrent cost of

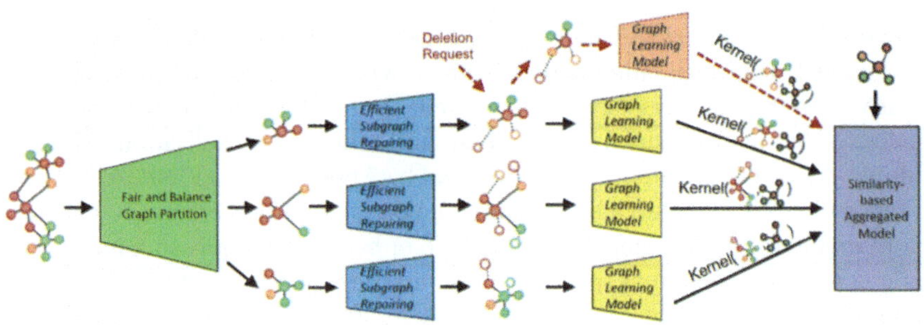

Fig. 5.5 Overview of GUIDE [9]. The training graph is partitioned into fair and balanced shards, each shard is locally repaired to compensate for cut edges, shard models are trained and only affected shards are updated upon deletion requests, and shard predictions are aggregated with weights determined by graph-kernel similarity to the test graph

retraining an aggregation model [9]. This design preserves the primary efficiency advantage of shard retraining while making each stage compatible with the inductive constraint.

Fair and Balanced Guided Partitioning To obtain predictable update cost and avoid label-skewed shards with weak local predictors, GUIDE formulates partitioning as constrained spectral clustering that enforces both balance (similar shard sizes) and fairness (each shard's label distribution approximates the global class proportions) [9]. Concretely, fairness and balance are encoded as linear constraints on the assignment matrix (or its relaxed embedding), yielding a constrained RatioCut objective that can be solved in a relaxed form and then discretised. The paper presents two instantiations: GPFB-Fast, which solves the relaxed constrained spectral problem and applies k-means to obtain fair/balanced shards, and GPFB-SR, which further introduces a spectral-rotation regulariser to better align the relaxed embedding with discrete partitions, improving partition quality at higher computational cost [9].

Efficient Subgraph Repairing for Utility Preservation Because partitioning necessarily removes cross-shard edges, nodes inside each shard lose a portion of their 1-hop neighbourhood, degrading message passing and harming utility. GUIDE mitigates this by generating synthetic neighbours within each shard to approximate the missing local context, using only local node features and degree information recorded before partitioning [9]. The authors propose three repair rules (Zero-Feature, Mirror-Feature, and MixUp-Feature neighbours) that create additional message-passing recipients but do not participate in the supervised loss, thereby improving representation propagation without introducing cross-shard data dependence.

Similarity-based Aggregation for Inductive Inference and Efficient Updates Finally, GUIDE replaces learning-based aggregation (which would require retraining aggregation weights after shard updates) with a graph-kernel similarity mechanism that independently assigns importance weights to shards based on their structural similarity to the test graph (Fig. 5.5, right) [9]. Using the pyramid match graph kernel, GUIDE computes a similarity score between the test graph and each shard's repaired subgraph, and aggregates shard predictions with weights proportional to these scores. Since each weight depends only on one shard and the current test graph, an unlearning request that updates a shard only requires recomputing that shard's similarity score, keeping the end-to-end update pipeline efficient under frequent deletions.

5.3 Summary

This chapter examined *graph unlearning* as the problem of updating graph learning pipelines so that a trained model behaves as if specified nodes, edges, attributes, or subgraphs had never been observed, under the distinctive complication of structural dependence induced by neighbourhood aggregation. We first formalised unlearning objectives for graph-structured data, distinguishing node-/edge-/subgraph-level deletions and clarifying the roles of transductive versus inductive regimes. We then organised the literature into a compact taxonomy (Table 5.1) spanning efficient retraining systems, certified approximate unlearning, and heuristic/editing-based methods.

Building on this taxonomy, we reviewed representative implementations across the spectrum. Retraining-based systems such as GraphEraser (Fig. 5.1) and GraphRevoker (Fig. 5.2) approximate retrain-on-modified-graph behaviour via partition–retrain–aggregate pipelines, with increasingly utility-aware sharding and aggregation. Certified approaches leverage convex surrogates and a unified Newton-style template (Fig. 5.3) to provide distributional closeness to retraining through residual bounds and calibrated randomness, as instantiated for transductive node tasks [4] and limited-data graph classification [5]. Finally, practical heuristic/editing-based methods target low-cost deployment, including training-free Fisher-guided parameter editing [7] and inductive shard-based unlearning with partition–repair–aggregate designs (Fig. 5.5) [9].

Overall, the chapter clarifies the trade space between guarantees, efficiency, and utility in graph unlearning, and shows why inductive constraints and propagation-driven non-locality often necessitate graph-specific redesign rather than direct reuse of i.i.d. unlearning techniques.

References

1. Min Chen, Zhikun Zhang, Tianhao Wang, Michael Backes, Mathias Humbert, and Yang Zhang. Graph unlearning. In *Proceedings of the 2022 ACM SIGSAC conference on computer and communications security*, pages 499–513, 2022.
2. Jiahao Zhang. Graph unlearning with efficient partial retraining. In *Companion Proceedings of the ACM Web Conference 2024*, pages 1218–1221, 2024.
3. Xiaoyu Xia, Ziqi Wang, Ruoxi Sun, Bowen Liu, Ibrahim Khalil, and Minhui Xue. Edge unlearning is not "on edge" ! an adaptive exact unlearning system on resource-constrained devices. In *2025 IEEE Symposium on Security and Privacy (SP)*, pages 2546–2563. IEEE, 2025.
4. Eli Chien, Chao Pan, and Olgica Milenkovic. Certified graph unlearning. *arXiv preprint arXiv:2206.09140*, 2022.
5. Chao Pan, Eli Chien, and Olgica Milenkovic. Unlearning graph classifiers with limited data resources. In *Proceedings of the ACM Web Conference 2023*, pages 716–726, 2023.
6. Jiancan Wu, Yi Yang, Yuchun Qian, Yongduo Sui, Xiang Wang, and Xiangnan He. Gif: A general graph unlearning strategy via influence function. In *Proceedings of the ACM Web Conference 2023*, pages 651–661, 2023.
7. Zhe-Rui Yang, Jindong Han, Chang-Dong Wang, and Hao Liu. Erase then rectify: A training-free parameter editing approach for cost-effective graph unlearning. In *Proceedings of the AAAI Conference on Artificial Intelligence*, volume 39, pages 13044–13051, 2025.
8. Chenhan Zhang, Weiqi Wang, Zhiyi Tian, and Shui Yu. Forgetting and remembering are both you need: Balanced graph structure unlearning. *IEEE Transactions on Information Forensics and Security*, 19:6751–6763, 2024.

9. Cheng-Long Wang, Mengdi Huai, and Di Wang. Inductive graph unlearning. In *32nd USENIX Security Symposium (USENIX Security 23)*, pages 3205–3222, 2023.
10. Mark EJ Newman. The structure and function of complex networks. *SIAM review*, 45(2):167–256, 2003.
11. Shoujin Wang, Liang Hu, Yan Wang, Xiangnan He, Quan Z Sheng, Mehmet A Orgun, Longbing Cao, Francesco Ricci, and Philip S Yu. Graph learning based recommender systems: A review. *arXiv preprint arXiv:2105.06339*, 2021.
12. Yonatan Harnik and Anat Milo. A focus on molecular representation learning for the prediction of chemical properties. *Chemical science*, 15(14):5052–5055, 2024.
13. Aidan Hogan, Eva Blomqvist, Michael Cochez, Claudia d'Amato, Gerard De Melo, Claudio Gutierrez, Sabrina Kirrane, José Emilio Labra Gayo, Roberto Navigli, Sebastian Neumaier, et al. Knowledge graphs. *ACM Computing Surveys (Csur)*, 54(4):1–37, 2021.
14. Jie Zhou, Ganqu Cui, Shengding Hu, Zhengyan Zhang, Cheng Yang, Zhiyuan Liu, Lifeng Wang, Changcheng Li, and Maosong Sun. Graph neural networks: A review of methods and applications. *AI open*, 1:57–81, 2020.
15. Lucas Bourtoule, Varun Chandrasekaran, Christopher A Choquette-Choo, Hengrui Jia, Adelin Travers, Baiwu Zhang, David Lie, and Nicolas Papernot. Machine unlearning. In *2021 IEEE Symposium on Security and Privacy (SP)*, pages 141–159. IEEE, 2021.
16. Chuan Guo, Tom Goldstein, Awni Y. Hannun, and Laurens van der Maaten. Certified data removal from machine learning models. In *Proceedings of the 37th International Conference on Machine Learning, ICML 2020, 13-18 July 2020, Virtual Event*, volume 119 of *Proceedings of Machine Learning Research*, pages 3832–3842. PMLR, 2020.

Chapter 6
Federated Unlearning

Having examined exact and approximate unlearning in centralized settings, auditing mechanisms for verifying unlearning guarantees, and recent developments in graph unlearning, this chapter turns to *federated unlearning*—the challenge of forgetting data in federated learning systems where training data remain distributed across clients.

In contrast to the centralized paradigm, federated learning coordinates model updates from multiple devices without collecting their raw data, which both heightens the importance of honouring deletion and withdrawal requests and complicates how such requests can be satisfied in practice. Federated unlearning must therefore address diverse granularities of forgetting (e.g., client-level, sample-level, or class-level removal), operate under communication and privacy constraints (e.g., secure aggregation and differential privacy), and remain robust in the presence of heterogeneous data, intermittent participation, and possibly malicious clients.

This chapter formalizes the federated unlearning problem and its threat models, surveys the main algorithmic strategies developed to approximate the effect of removing clients or data without full retraining, and discusses evaluation protocols and open challenges in making federated unlearning practical and trustworthy in real-world deployments.

6.1 What Is Federated Unlearning

Federated learning (FL) is a distributed training paradigm in which multiple clients collaboratively learn a shared model while keeping their raw data local, thereby reducing direct data centralisation and supporting privacy- and regulation-aware learning at scale [1]. As illustrated in Fig. 6.1, FL proceeds in rounds: the central server broadcasts the current global model to selected clients, each client performs local model training on its private dataset, and only model updates are sent back for

W. Wang, S. Yu, *Machine Unlearning: Concepts and Implementations*,
Digital Privacy and Security, https://doi.org/10.1007/978-981-92-1142-5_6

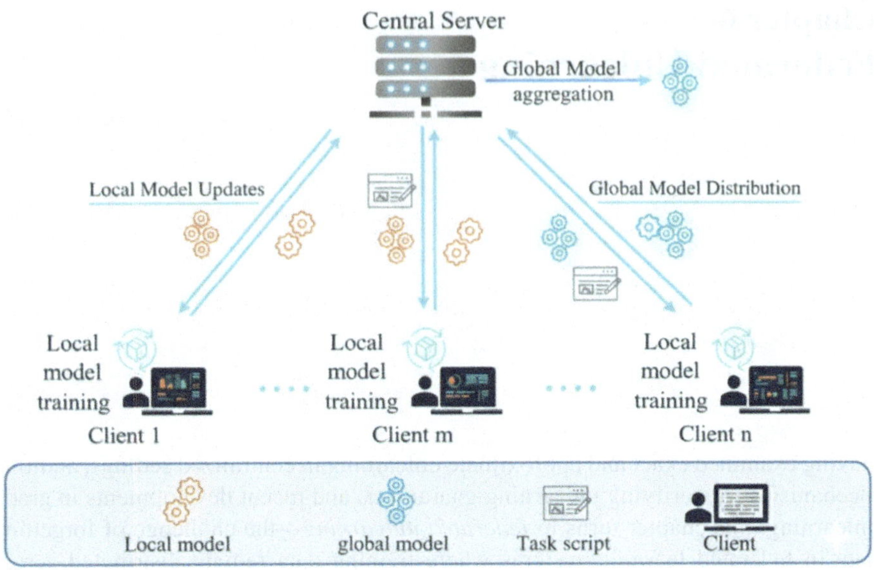

Fig. 6.1 A typical federated learning framework

global model aggregation, after which the updated model is redistributed for the next round [1]. In practice, FL must contend with system and statistical heterogeneity and, critically, communication bottlenecks, motivating communication-efficient designs such as quantised and lazily aggregated update schemes [2]; moreover, although FL avoids sharing raw data, it does not automatically guarantee privacy against inference from updates, which has led to complementary privacy-preserving computation techniques (e.g., secure aggregation, differential privacy, and cryptographic protocols) to strengthen confidentiality guarantees in real deployments [3].

Federated unlearning (**FU**) extends machine unlearning to the federated learning settings, aiming to remove the influence of specific clients, user groups, or data subsets from a collaboratively trained global model, while respecting core FL constraints such as privacy, limited data accessibility, and heterogeneous data/systems. Given a global model produced by aggregating local client updates, FU aims to output an updated model that behaves (approximately) as if the target client(s) or data had never participated in training, but without retraining from scratch across all remaining clients. Compared with centralized unlearning, FU is complicated by

- (i) where the information needed for removal resides (often only on-device),
- (ii) what is practically storable at the server (e.g., checkpoints or update logs),
- and (iii) how to restore utility under non-i.i.d. data and partial participation.

To clarify the design space, we organise FU methods along three largely orthogonal axes: *(1) Unlearning granularity* (client/user-level vs. class-/group-/sample-level), *(2) Execution locus* (client-side, server-side, or hybrid), and *(3)*

Table 6.1 The design-space of federated unlearning along key design axes

Axis	Typical options and representative examples
Unlearning granularity	Client/user-level contribution removal [4–6]; class-/group-wise forgetting [7].
Execution locus	Hybrid client untraining + limited global rounds [4]; purely server-side rollback/compensation [5]; requesting-client-only approximate cancellation [8].
History dependence	Log/checkpoint-based rollback and fast reconstruction [6]; no server-side history via client-side influence approximation [8]; privacy-preserving reuse of historical information [9].

Reliance on historical information (no history, lightweight summaries, or stored per-client update logs/checkpoints). The design-space framework is summarized in Table 6.1, which useful because many FU proposals differ less in their high-level goal than in which party performs the removal, what auxiliary information is assumed available, and what level of assurance is targeted.

Client/User-level Removal via Rollback or Compensation Early FU work operationalises client-level forgetting by explicitly reversing or compensating for training dynamics under limited additional communication. Halimi et al. remove a client by locally "untraining" the target client through constrained loss maximisation and then running only a small number of additional global rounds with the remaining clients to restore utility, thereby approaching retraining-level performance without storing raw data or detailed update histories [4]. Wu et al. instead propose a purely server-side procedure that subtracts the accumulated historical updates of the removed client and then uses knowledge distillation on an auxiliary dataset to recover accuracy, avoiding any further client participation [5]. In personalised on-device recommendation, Yuan et al. frame unlearning as a log-based rollback problem (FRU): the server rolls back and calibrates stored, compressed, and importance-filtered parameter updates associated with a user, and reuses them to accelerate reconstruction of the recommender after forgetting [6]. These methods primarily differ in where the cancellation happens (client vs. server) and what is assumed available (fresh client participation vs. stored histories).

Beyond Client-level: Class-/group-/sample-targeted FU and Objective-integrated Designs FU has expanded beyond removing entire clients to more targeted forms of forgetting that better match application needs. For instance, Wang et al. propose class-discriminative pruning that estimates how strongly channels support particular classes and prunes those associated with classes to be forgotten, followed by fine-tuning on retained data, enabling class-wise unlearning without direct access to raw client data [7]. A second line of work builds unlearning directly into the training objective or the update rule. BFU frames unlearning as variational Bayesian inference, designing updates that forget the erased samples while preserving performance on the remaining data under federated aggregation [10]. FedU removes the need for the server to store clients' training history by having only the clients who request unlearning compute an influence-based approximation on their remaining local data to approximately subtract (cancel) their past contribution, while all other clients continue standard FL training [8]. Collectively, these approaches illustrate a shift from explicit rollback toward model-centric (pruning/editing) and objective-centric (Bayesian/influence) unlearning.

System, Protocol, and Recoverability Considerations Complementary work focuses on making FU deployable and trustworthy under realistic system conditions. Su and Li extend FU to asynchronous FL by clustering clients so that unlearning and retraining can proceed asynchronously, reducing straggler-induced delays [11]. Lin et al. study incentive and client-selection mechanisms (e.g., reinforcement-learning-based policies) to encourage rational clients to participate in costly unlearning rounds under limited budgets [12]. Liu et al. design privacy-preserving protocols (e.g., two-party computation) that enable reuse of historical client information while providing certified bounds on the deviation between the unlearned and fully retrained models, directly addressing leakage risks from storing or manipulating past updates [9]. More recently, "recoverable" FU recognises that users may temporarily leave and later rejoin: RFUL couples a knowledge-unlearning phase with a fast knowledge-recovery phase, enabling efficient removal and later restoration using only the target client's data/model without requiring other clients to participate [13].

Across this taxonomy, FU methods trade off communication and computation cost, assumptions about stored history or auxiliary data, and strength of guarantees. The remainder of this chapter builds on these axes to compare FU designs and motivate recoverable unlearning as a practical mechanism for dynamic participation and regulatory compliance.

6.2 Federated Unlearning Implementation

6.2.1 Client-Level Federated Unlearning

Efficiently Unlearn a Client in Federated Learning Halimi et al. study client-level federated unlearning in cross-silo FL, where one organization withdraws after standard Federated Averaging (FedAvg) training and requests removal of its entire contribution [4]. Their goal is to approximate the model that would have been obtained by retraining without the target client, but without requiring the server (or clients) to store per-round client updates or global checkpoints. To this end, they propose a two-stage procedure: (i) the departing client performs a local "scrubbing" step that reverses its own training signal, and (ii) the server and remaining clients run only a small number of additional FL rounds to restore global utility.

The key technical component is the local unlearning step, cast as a constrained loss maximization problem solved by projected gradient descent (i.e., projected gradient ascent on the local empirical loss) [4]. Rather than naively maximizing an unbounded loss (which can quickly drive the parameters toward a degenerate solution), the target client maximizes its local objective while restricting the parameters to an ℓ_2 ball around a reference model θ_{ref} intended to preserve knowledge from the retained clients. Concretely, θ_{ref} is chosen as the average of the other clients' last-round local models, and—under FedAvg with always-participating clients—the target client can compute it from the final global model θ^T and its own last local model θ_i^{T-1} via

$$\theta_{\text{ref}} = \frac{1}{N-1}\sum_{j\neq i}\theta_j^{T-1} = \frac{N\theta^T - \theta_i^{T-1}}{N-1}. \tag{6.1}$$

Unlearning then solves $\max_{\|\theta-\theta_{\text{ref}}\|_2\leq\delta} F_i(\theta)$ using projected updates, with early stopping (e.g., a distance-based threshold) to control how far the scrubbed model drifts from the pre-unlearning local solution.

Starting from this locally unlearned model, the post-training phase runs a few additional FedAvg rounds using only the retained clients to re-align the global parameters with the retained data distribution, improving fidelity while maintaining low performance on the erased client's distribution [4]. Experiments on MNIST, EMNIST, and CIFAR-10 consider settings that deliberately separate the target client's distribution (e.g., backdoor-triggered or "flipped" data) and evaluate three criteria—efficacy (forgetting on the target distribution), fidelity (utility on retained data), and efficiency (communication cost versus full retraining)—showing that brief post-training can recover near-retraining utility at substantially lower communication cost (reported reductions on the order of 5×–24×).

Federated Unlearning with Knowledge Distillation Wu et al. propose a server-side client-level unlearning pipeline that removes a target client's influence without further client participation by (i) subtracting the client's stored historical updates from the final global model and (ii) repairing the resulting model via knowledge distillation using an auxiliary dataset available at the server [5]. They motivate this setting by three practical constraints in FL: later-round updates depend on earlier-round states, training is stochastic due to client sampling and local SGD, and the server does not access clients' raw data. To make unlearning measurable, they embed a backdoor into the target client's training process and assess forgetting by how much the backdoor attack success rate drops, while tracking clean-test accuracy to quantify utility preservation [5].

Update Subtraction In a FedAvg view, the final global parameters can be written as the initial model plus accumulated round-wise updates, where each round update decomposes into per-client contributions [5]. Let θ^0 be the initial model and θ^T the final model after T rounds. For round t, with selected clients S_t, define each participating client's update Δ_k^t and the aggregated update

$$\Delta^t = \frac{1}{|S_t|}\sum_{k\in S_t}\Delta_k^t, \qquad \theta^T = \theta^0 + \sum_{t=0}^{T-1}\Delta^t. \tag{6.2}$$

Exact "remove-client-i then retrain" would, in principle, require recomputing the entire trajectory because removing Δ_i^t changes subsequent models and thus subsequent updates. Wu et al. instead introduce a lazy approximation that keeps the original trajectory fixed and simply zeros out the target client's contributions at each round, yielding a closed-form "subtracted" model:

$$\tilde{\theta}^{-i} = \theta^T - \sum_{t=0}^{T-1} \mathbf{1}[i \in S_t] \frac{1}{|S_t|} \Delta_i^t. \tag{6.3}$$

This produces a fast, communication-free unlearning model, but it can be skewed, i.e., accuracy may drop due to distributional drift and the fact that subsequent rounds were trained on states that implicitly depended on the removed client [5].

Server-only Distillation Repair To mitigate this skew, they "repair" $\tilde{\theta}^{-i}$ via knowledge distillation on an outsourced dataset at the server, treating the original global model θ^T as a teacher and the subtracted model $\tilde{\theta}^{-i}$ as a student [5]. With unlabeled auxiliary inputs x, the teacher provides soft targets using a temperature $\tau > 1$,

$$p_T(\cdot|x) = \text{softmax}(z_T(x)/\tau), \qquad p_S(\cdot|x) = \text{softmax}(z_S(x)/\tau), \tag{6.4}$$

and the student is trained to match the teacher (e.g., by minimizing $\text{KL}(p_T \parallel p_S)$); when labels are available, a weighted combination of hard-label loss and distillation loss can be used, following standard practice [5]. Conceptually, this step restores much of the lost utility without requiring any client data or additional FL rounds and helps avoid re-imprinting the target client's backdoor behavior.

Federated Unlearning with Fast Recovery Zhou et al. [13] propose RFUL, a recoverable federated unlearning framework that supports both (i) fast removal of a user's contribution from a federated model and (ii) fast re-integration of that contribution when the same user later returns, under a stringent setting where only the requesting user's data are available for both operations. Motivated by practical class-level scenarios (e.g., federated face recognition where each user corresponds to a class), RFUL goes beyond prior federated unlearning methods that treat forgetting as permanent and often rely on non-target users' data to prevent catastrophic forgetting. Instead, RFUL explicitly couples knowledge unlearning with knowledge recovery, enabling users to flexibly withdraw from and rejoin the global model without requiring access to the preserved data.

OVR Architecture for Localized Transfer As illustrated in the top panel of Fig. 6.2, RFUL adopts a one-vs-rest (OVR) design that factorizes the model into a shared feature extractor and K independent OVR heads (sigmoid-based nodes) that each distinguish one class from the rest. This multi-head structure reduces cross-class interference during unlearning/recovery: operations can target the OVR node(s) associated with the requested class(es), while the classifier parameter space for preserved classes can be protected by freezing their corresponding heads.

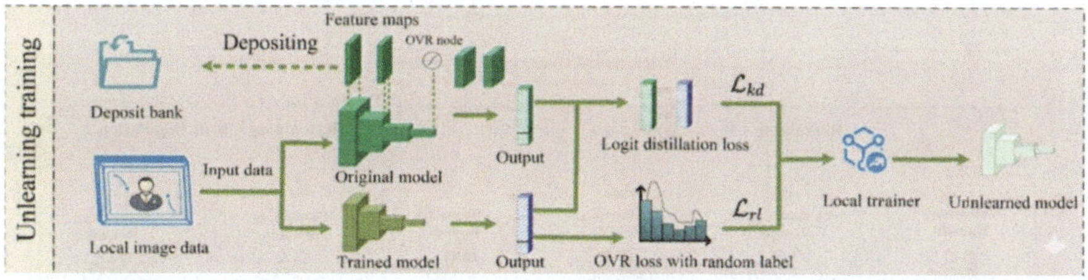

Fig. 6.2 Training workflows in RFUL [13]: knowledge unlearning that deposits feature maps/OVR nodes and updates the model using random-label OVR loss L_{rl} plus logit distillation L_{kd} to produce M_1

Knowledge Unlearning: Random Relabeling with Logit Distillation Regularization The middle panel of Fig. 6.2 depicts RFUL's unlearning stage. Starting from the original global model M_0, the target client first deposits auxiliary information for possible future recovery (selected intermediate feature maps of its local samples and the relevant OVR node parameters) into a local deposit bank. Unlearning then degrades the model's discriminative ability on the target data D_f by assigning random labels to D_f and optimizing an OVR loss on this mislabeled set. They freeze the OVR heads associated with preserved classes to avoid direct classifier-space drift on non-target categories. Because training only on D_f can still perturb the shared feature extractor and induce catastrophic forgetting, RFUL adds a logit distillation term L_{kd} that matches the current model's output scores to those produced by the original model M_0 on the same target inputs, acting as a stabilizing regularizer when D_r is unavailable. The combined objective produces an unlearned model M_1 whose predictions on D_f become near-random while largely preserving performance on non-target classes.

Knowledge Recovery: Feature Distillation with Classifier Replacement and Freezing The bottom panel of Fig. 6.2 shows RFUL's recovery stage. To restore previously removed contributions using only D_f, the client begins from the unlearned model M_1 but replaces the target-class OVR node(s) with the stored ones from the deposit bank and then freezes classifier parameters. Recovery updates are therefore concentrated in the shared feature extractor and are guided by (i) the standard OVR classification loss on D_f (term L_{ovr}) and (ii) a feature distillation loss L_{fk} that drives the current feature extractor to reproduce the deposited intermediate feature maps. This feature-level constraint steers optimization back toward the original representational geometry without accessing D_r, enabling RFUL to recover a model M_2 that performs well on both the recovered target classes and the preserved classes.

Incentive and Dynamic Client Selection for Federated Unlearning Lin et al. investigate federated unlearning in a hierarchical, cluster-based FL architecture and ask a practical question: when some clients request to be forgotten, how can the system (i) motivate the remaining clients to participate in the extra unlearning workload and (ii) schedule those clients in a data-aware way under non-IID distributions and multiple concurrent unlearning tasks [12].

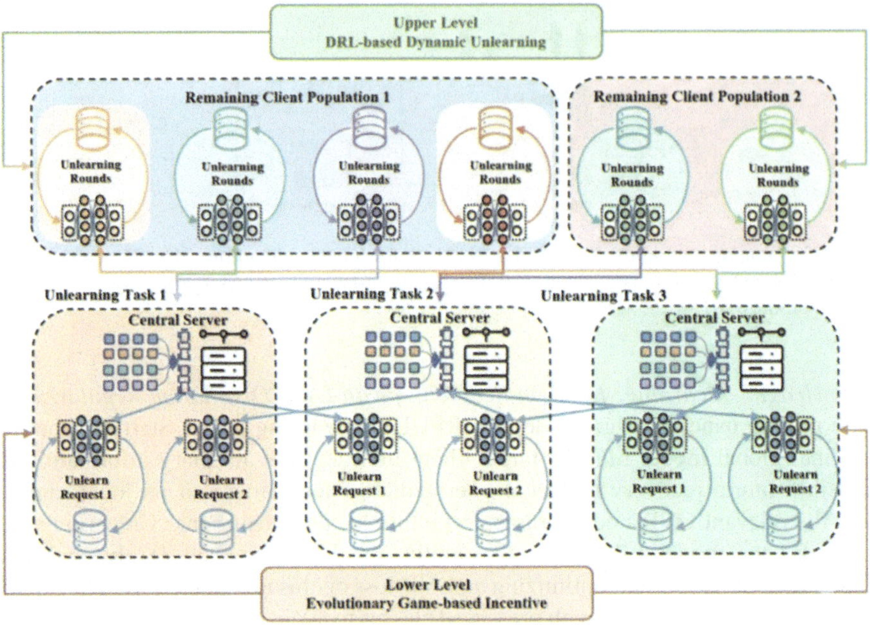

Fig. 6.3 Two-level framework for hierarchical federated unlearning in dynamic environments [12]. The lower level models incentive-compatible participation of remaining clients across cluster heads via an evolutionary game, while the upper level uses DRL to dynamically schedule a subset of remaining clients for each unlearning round under non-IID data and multiple concurrent unlearning tasks

As illustrated in Fig. 6.3, the system comprises multiple cluster heads (central servers) that coordinate unlearning for different AIGC tasks, while unlearning requests arrive over time and the set of active clients changes across rounds. The authors observe that many client-level federated unlearning pipelines implicitly assume that remaining clients will participate "for free" and can be selected randomly or statically, which overlooks their added communication, computation, and storage burdens and can be inefficient or unstable when data are highly heterogeneous. To address this, they propose a two-level orchestration framework (Fig. 6.3): a lower-level evolutionary game that aligns participation incentives across clusters, and an upper-level DRL scheduler that dynamically selects a subset of remaining clients per unlearning round.

At the lower level, Lin et al. formulate an incentive mechanism using evolutionary game theory to capture how heterogeneous groups of remaining clients adapt their participation decisions in response to rewards and costs [12]. Clients are grouped into populations according to their data volume and resource profiles, and each population distributes its participation across cluster heads through an association-probability strategy. For each population–server pair, the utility is defined as allocated reward minus participation costs, including communication, computation, and storage overheads incurred by local calibration and maintaining historical information required by the unlearning process. Population strategies evolve according to replicator dynamics, so association probabilities increase for server choices that yield above-average utility and decrease otherwise. The resulting equilibrium characterizes a stable, incentive-consistent allocation of remaining clients across clusters, which improves resource efficiency and mitigates imbalance effects under non-IID data, as conceptually depicted by the bottom layer in Fig. 6.3.

At the upper level, the authors introduce a DRL-based dynamic client selection module (double DQN) that, within each unlearning round, schedules a subset of remaining clients whose participation is expected to best support recovery of the unlearned global model under heterogeneity [12]. As shown at the top of Fig. 6.3, the scheduler operates in a dynamic environment where client states and task conditions vary across rounds. The DQN state encodes each client's alignment with the current unlearning trajectory using cosine similarities between its calibrated update and the current global unlearning model, thereby injecting data-distribution signals into scheduling. The action is a (potentially large) binary selection over remaining clients, and the reward is shaped around accuracy recovery toward a target (evaluated on server-side validation data), with additional terms encouraging faster convergence. By coupling incentive-compatible participation (lower level) with data-aware, round-by-round scheduling (upper level), the framework targets federated unlearning that is both economically sustainable for remaining clients and more robust under non-IID data than static or random client-selection baselines [12].

6.2.2 Class and Sample-Level Federated Unlearning

Federated Unlearning via Class-Discriminative Pruning (FUCP) Wang et al. study category-level federated unlearning for CNN image classifiers, where a trained global model must remove information about one (or several) sensitive classes after training, while the federated server has no direct access to the decentralized raw data [7]. Their key observation is that convolutional channels contribute unevenly across classes: some channels are highly specialized to particular categories, whereas others encode more shared features. FUCP operationalizes this insight by identifying channels that are most discriminative for the class to be forgotten and then erasing class-specific capacity through structured (channel-wise) pruning, followed by a short federated fine-tuning stage to recover utility on retained classes.

Client-to-server Statistics: From Image-wise Activations to Global Class-wise Activations As illustrated in Fig. 6.4, FUCP begins by distributing an unlearning program to participating clients, who use their locally trained models to transform private images into lightweight channel–class statistics rather than sharing data [7]. Concretely, each client feeds labeled images through the CNN and converts feature

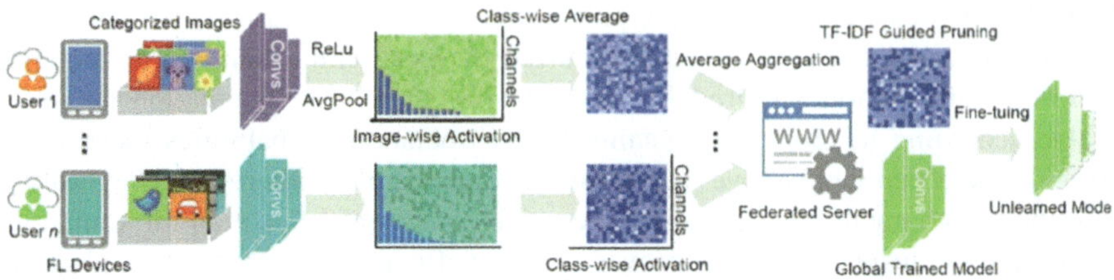

Fig. 6.4 Workflow of Federated Unlearning via Class-Discriminative Pruning [7]. Clients compute class-wise channel-activation statistics from local images (ReLU + AvgPool → image-wise activation → class-wise average). The server aggregates these statistics and computes TF–IDF-based class-discriminative scores for channels, then performs one-shot channel pruning for the target class and triggers federated fine-tuning on target-excluded data to obtain the unlearned model

maps into per-channel activation scores (ReLU followed by spatial average pooling), producing an image-wise activation vector. These vectors are then averaged within each class to obtain a class-wise activation profile per channel. The server aggregates the uploaded client profiles (via averaging) into a global class-wise activation representation, which serves as a proxy for how strongly each channel supports each class under federated non-IID data.

TF–IDF-guided Channel Scoring and One-shot Pruning FUCP scores class discrimination with a TF–IDF-style metric by treating the output of a channel as a "word" and the feature-map representation of a class as a "document" [7]. The term-frequency component prioritizes channels that respond strongly and frequently to the target class, while the inverse-document-frequency component downweights channels whose activations are common across many classes, so that the remaining high-score channels are those that are both salient and class-specific. Using the resulting TF–IDF scores, the server constructs a pruner and performs one-shot channel pruning: channels whose scores fall into a pre-defined top proportion are removed (and the dependent filters are pruned accordingly), yielding a thinner global model with reduced representational capacity for the forgotten class.

Federated fine-tuning on target-excluded data and empirical efficiency After pruning, FUCP restores overall accuracy by running standard federated training for a short period on target-excluded local data (i.e., clients continue training but filter out the class to be forgotten), without introducing additional regularizers beyond normal FL optimization [7]. Empirically, this prune-once-then-fine-tune recipe substantially reduces unlearning time compared with retraining from scratch while maintaining retained-class accuracy: on CIFAR-10, the reported unlearning speedups are 8.9× (ResNet) and 7.9× (VGG), and on CIFAR-100 they are 9.9× (ResNet) and 8.4× (VGG).

Bayesian Federated Unlearning with Parameter Self-Sharing Wang et al. propose *Bayesian Federated Unlearning* (BFU), which casts federated unlearning as approximate Bayesian posterior updating within a variational Bayesian federated learning (BFL) pipeline [10]. In their setting, the global model is a Bayesian neural network (BNN) represented by a variational posterior maintained at the server and updated by aggregating client-side approximate posteriors in a FedAvg-style manner. When a client receives a deletion request, it partitions its local dataset into an erased subset D_k^e and a remaining subset D_k^r, performs unlearning locally using only D_k^e, and then uploads an updated posterior to the server for standard aggregation. This design avoids sharing raw data and removes the need for the server to store historical client updates, while allowing unlearning to be executed as an in-protocol client operation rather than a server-orchestrated retraining procedure [10].

At the method level, BFU targets the ideal posterior that would be obtained by retraining on the remaining data only, and approximates this target by minimizing a KL divergence between an updated variational posterior and the (intractable) post-erasure posterior. The resulting client-side objective decomposes into (i) a term that drives forgetting by reducing the likelihood of the erased samples under the updated posterior and (ii) a regularization term that constrains the updated posterior to stay close to the pre-unlearning global posterior. To stabilize this trade-off, BFU introduces an unlearning rate β that scales the erased-data likelihood term, so that forgetting pressure can be tuned to the erasure ratio and task difficulty, mitigating catastrophic unlearning in which utility collapses on benign data [10]. Practically, an unlearning client performs a short local optimization loop (SGD on the variational parameters) using minibatches from D_k^e, and employs a simple verification rule that checks whether the model's confidence on erased samples drops to (or below) a random-guess baseline before terminating local unlearning and returning the updated posterior for aggregation [10].

To further reduce utility degradation when $|D_k^e|$ is large or the task is complex, the authors propose BFU-SS, a parameter self-sharing variant inspired by hard parameter sharing in multi-task learning [10]. BFU-SS treats unlearning and retention as two coupled objectives optimized jointly on the same Bayesian model: within each local epoch, the client samples one minibatch from the erased set to compute the BFU unlearning loss and another minibatch from the remaining set to compute the standard BFL training loss. The client then updates a single shared posterior using a weighted sum of these two losses, which explicitly regularizes the unlearning trajectory by continued learning signals from D_k^r. By coupling forgetting with retention during the same local optimization phase, BFU-SS empirically improves the stability of unlearning and preserves accuracy more effectively than applying BFU as an isolated erased-data optimization step [10].

Federated Unlearning via User-Side Influence Approximation Forgetting Wang et al. propose FedU, a federated unlearning method that localizes the unlearning workload to the requesting user while leaving the federated learning (FL) protocol unchanged for the server and non-requesting users [8], as illustrated in Fig. 6.5. Consider a user C_{k_u} who wishes to delete a subset of local samples. FedU requires C_{k_u} to partition its local dataset into an erased set $D_{k_u}^e$ and a remaining set $D_{k_u}^r$, then perform additional local updates that remove the influence of $D_{k_u}^e$ before resuming ordinary FL participation. The unlearning goal follows the standard

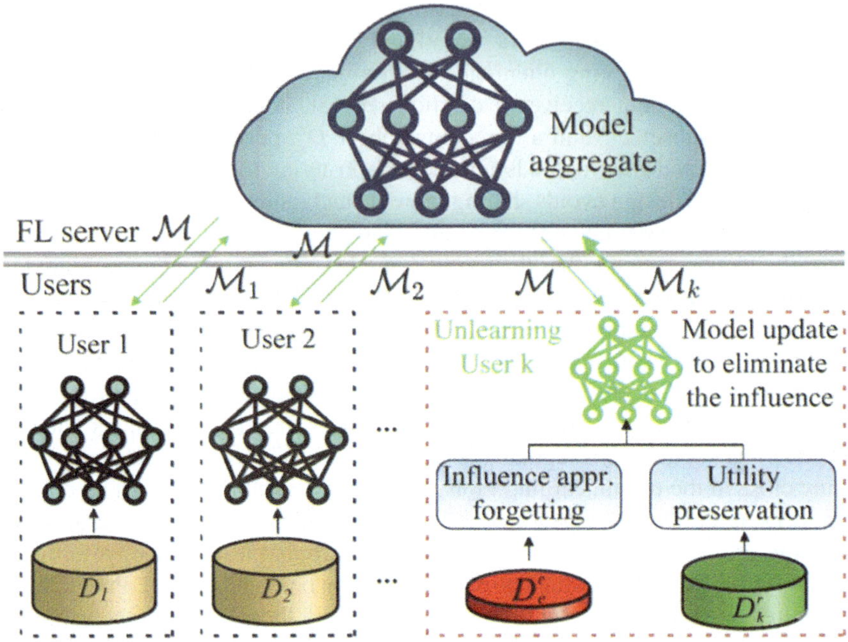

Fig. 6.5 Overview of FedU [8]. Only the requesting user performs additional local unlearning steps. The user splits its local dataset into an erased subset D_k^e and a remaining subset D_k^r, then applies influence approximation forgetting (IAF) to remove the effect of D_k^e and utility preservation (UP) updates on D_k^r to stabilize performance. The server and non-requesting users continue standard FL training and aggregation

retraining-equivalence ideal: the resulting unlearned global model should match the model distribution obtained by rerunning the original FL algorithm on the residual global dataset in which $D_{k_u}^e$ has been removed.

The key mechanism is *influence approximation forgetting* (IAF), which performs data erasure by subtracting an estimated parameter influence term induced by the erased samples [8]. Starting from the trained model parameters θ^*, FedU approximates the retrained parameters using a first-order influence-function update of the form "inverse Hessian times erased-gradient sum". Since computing a global Hessian is infeasible in FL, the requesting user estimates the required inverse-Hessian–vector product locally via stochastic Hessian–vector products computed on samples drawn from $D_{k_u}^r$. This yields an efficient, fully client-side approximation of the parameter change that would occur if $D_{k_u}^e$ had never been used, without accessing other users' data or storing per-round server-side histories.

Because an influence-based update can over-correct and degrade accuracy, FedU further introduces a *utility preservation* (UP) component and an adaptive update rule, yielding the full method FedU-U [8]. UP continues optimizing the unlearned local model on a small minibatch sampled from $D_{k_u}^r$ using the original FL training loss, which stabilizes performance on benign data. FedU-U then treats erasure and utility as a two-objective problem and computes a single update direction that jointly accounts for the gradients of the IAF loss and the UP loss, reducing sensitivity to manual trade-off tuning. After each communication round, the requesting user applies a verification test on $D^e_{k_u}$; once the test is satisfied, the user returns to normal FL training.

Asynchronous Federated Unlearning Su and Li propose *KNOT*, a clustered aggregation strategy for asynchronous federated learning (FL) that reduces the wall-clock overhead of retraining after data-erasure requests [11]. Their starting point is that unlearning becomes costly in standard FL because influence quickly turns global: repeated server-wide aggregation causes one client's updates (and hence its data influence) to diffuse into the model states of many other clients. Consequently, even if the erasure request originates from a single client, retraining-based unlearning tends to interrupt a large fraction of the system, while approximation-based methods (e.g., FedEraser) still require broad participation and typically provide only approximate removal. KNOT instead exploits the inherent concurrency of asynchronous FL by restructuring aggregation so that influence propagation is localized. As illustrated in Fig. 6.6, clients are partitioned into clusters and aggregated asynchronously within each cluster; when an erasure request occurs in one cluster, only that affected cluster rolls back and retrains from scratch, whereas other clusters continue to send updates and make forward progress. This "containment" effect decouples unlearning latency from the training progress of the entire system [11].

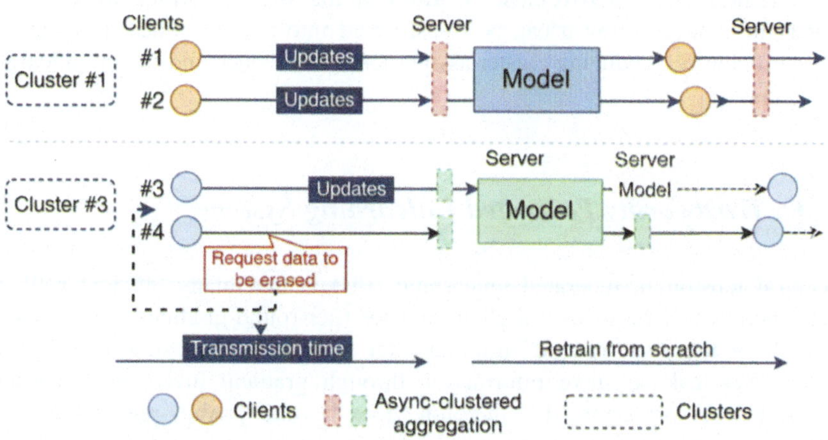

Fig. 6.6 Asynchronous clustered aggregation for federated unlearning [11]. Clients are partitioned into clusters and aggregated asynchronously within each cluster. Upon a data-erasure request in one cluster, only that cluster rolls back and retrains, while other clusters continue standard asynchronous training

The method is developed in two stages. First, the authors introduce random clustered aggregation, which assigns clients to a small number of clusters and restricts model mixing to intra-cluster aggregation, acting as a firewall that limits cross-client influence diffusion [11]. Asynchronous operation is essential here: because clusters progress at different speeds, a rollback triggered by an erasure request need not stall the whole system. Only the impacted cluster retrains, while other clusters proceed with their own asynchronous rounds (Fig. 6.6). The clustered design is also compatible with different retraining procedures, since it modifies the aggregation topology rather than the local optimizer. To determine when to stop cluster-local training and merge clusters, the authors propose simple termination rules based on cluster-wise validation accuracy and its variance, followed by a final cross-cluster aggregation. On MNIST, this random clustering already reduces unlearning-induced slowdown compared with synchronous FedAvg and asynchronous FedBuff, because the performance disruption is confined to a subset of clusters rather than propagating system-wide [11].

Building on this baseline, KNOT replaces random clustering with an optimization-driven client-to-cluster assignment that explicitly accounts for both system heterogeneity and statistical heterogeneity [11]. The key intuition is to form clusters that are simultaneously fast and statistically coherent: clients with shorter local training times and smaller update disparity relative to the global trajectory yield clusters that converge more quickly and are less affected by non-IID drift. Concretely, KNOT defines a per-client training time T_k and a model-disparity score S_k (measured via cosine similarity between a client's update and the global update), sets anchor targets for each cluster, and then solves a lexicographic minimization problem over binary assignment variables to balance computational load and representation quality across clusters. Although the original formulation is integer-valued, the authors show it can be transformed into a single linear program and solved efficiently, producing cluster assignments that reduce end-to-end unlearning time [11].

6.2.3 Trustworthy Federated Unlearning Systems

A central tension in federated unlearning (FU) is that many efficient rollback-style schemes rely on historical client updates (per-round gradients, local models, or auxiliary logs) to "rewind" training after a deletion request, yet these same artifacts can leak sensitive information through gradient inversion and related attacks. Privacy-preserving FU therefore goes beyond privacy-preserving FL: it must protect client information not only during routine training-time aggregation, but also during long-horizon storage and reuse of historical signals for unlearning.

Liu et al. study this problem under dynamic user participation and propose to couple clustering-based FU with secure aggregation so that unlearning can be localized while individual gradients remain hidden [14]. Their workflow groups users into clusters, runs FL within each cluster using SecAgg+ for privacy-preserving aggregation, and performs inference by ensembling cluster models; crucially, when an unlearning request removes a target user, only the cluster containing that user needs to be retrained, which limits the blast radius of rollback and reduces system-wide overhead [14]. The core technical challenge is that SecAgg+ security depends on cluster composition: dropouts and unlearned users shrink clusters, and adversarial users can collude to break Shamir-based protections. To address this, the authors formalize clustering requirements that ensure (i) adversarial users per cluster stay below the Shamir threshold and (ii) enough honest users remain to reconstruct masked secrets despite dropouts and removals, and they design clustering procedures that maintain SecAgg+ compatibility as users join, drop, and are deleted [14].

Complementing this line, Liu et al. propose Starfish, which targets privacy leakage from server-side FU history by executing rollback-style unlearning under two-party computation (2PC) while still supporting standard FL training [9]. Starfish adopts a two-server architecture: clients additively secret-share their per-round gradients and auxiliary values across two non-colluding servers, which then perform secure computation to obtain the aggregated update while never revealing any single client's raw gradients to either server [9]. When a client-removal request arrives, Starfish reduces secure-computation cost by selecting only the most influential historical rounds (via a secure round-selection protocol) [9], then reconstructs an approximate unlearning trajectory using an L-BFGS-style update estimation procedure, and periodically performs privacy-preserving threshold checks to correct accumulated approximation error [9]. Beyond empirical unlearning quality, Starfish further provides a certified client-removal bound that upper-bounds the deviation between its unlearned model and retraining-from-scratch under standard smoothness and strong convexity assumptions [9].

Taken together, these two works illustrate two complementary design patterns for privacy-preserving FU. The dynamic-participation scheme achieves privacy primarily through SecAgg+-compatible clustering and localizes retraining to affected clusters [14], whereas Starfish protects the FU-critical history itself via secret-sharing plus 2PC to enable server-side rollback without exposing stored gradients [9]. Both are fundamentally client-level FU in their target (removing a user's contribution), but they differ in system assumptions (single-server SecAgg inside clusters versus two non-colluding servers) and in where the privacy cost is paid (cluster maintenance versus secure history management).

6.3 Summary

This chapter reviewed federated unlearning (FU), the problem of removing the influence of specified clients or data subsets from a collaboratively trained federated model under privacy, communication, and heterogeneity constraints. We first formalized FU and organized the design space along three axes: unlearning granularity (client- vs. class/group/sample-level), execution locus (client-side, server-side, or hybrid), and reliance on historical information (none, lightweight summaries, or stored logs/checkpoints), providing a unified lens for comparing seemingly diverse methods.

We then surveyed representative FU strategies. For client/user-level removal, we covered hybrid client scrubbing with brief post-training to recover utility, server-side update subtraction with distillation-based repair, and recoverable FU that supports both fast forgetting and later re-integration. For more targeted forgetting, we discussed class-discriminative pruning and objective-integrated approaches such as Bayesian FU and user-side influence approximation, which embed forgetting into local optimization without extensive server-side histories. We further examined systems-level mechanisms that improve deployability under realistic conditions, including asynchronous clustered aggregation to localize rollback costs and incentive-aware, data-aware client selection to manage participation and non-IID effects.

Finally, we highlighted trustworthy FU as a core requirement for deployment. Privacy-preserving FU must protect not only training-time updates but also the long-horizon historical information used for unlearning. We reviewed secure-aggregation-compatible clustering under dynamic participation and 2PC-based rollback with certified client removal as two complementary pathways toward privacy-preserving and verifiable unlearning. Overall, FU methods exhibit clear trade-offs between efficiency, assumptions on stored history or auxiliary resources, and the strength of assurance, motivating future work on robust, privacy-preserving, and auditable FU in dynamic real-world federated systems.

References

1. Chen Zhang, Yu Xie, Hang Bai, Bin Yu, Weihong Li, and Yuan Gao. A survey on federated learning. *Knowledge-Based Systems*, 216:106775, 2021.
2. Jun Sun, Tianyi Chen, Georgios B Giannakis, Qinmin Yang, and Zaiyue Yang. Lazily aggregated quantized gradient innovation for communication-efficient federated learning. *IEEE Transactions on Pattern Analysis and Machine Intelligence*, 44(4):2031–2044, 2020.
3. Jingxue Chen, Hang Yan, Zhiyuan Liu, Min Zhang, Hu Xiong, and Shui Yu. When federated learning meets privacy-preserving computation. *ACM Computing Surveys*, 56(12):1–36, 2024.
4. Anisa Halimi, Swanand Kadhe, Ambrish Rawat, and Nathalie Baracaldo. Federated unlearning: How to efficiently erase a client in fl? *arXiv preprint arXiv:2207.05521*, 2022.
5. Chen Wu, Sencun Zhu, and Prasenjit Mitra. Federated unlearning with knowledge distillation. *arXiv preprint arXiv:2201.09441*, 2022.
6. Wei Yuan, Hongzhi Yin, Fangzhao Wu, Shijie Zhang, Tieke He, and Hao Wang. Federated unlearning for on-device recommendation. In *Proceedings of the sixteenth ACM international conference on web search and data mining*, pages 393–401, 2023.
7. Junxiao Wang, Song Guo, Xin Xie, and Heng Qi. Federated unlearning via class-discriminative pruning. In *Proceedings of the ACM Web Conference 2022*, pages 622–632, 2022.
8. Weiqi Wang, Chenhan Zhang, Zhiyi Tian, and Shui Yu. Fedu: Federated unlearning via user-side influence approximation forgetting. *IEEE Transactions on Dependable and Secure Computing*, 2024.

9. Ziyao Liu, Huanyi Ye, Yu Jiang, Jiyuan Shen, Jiale Guo, Ivan Tjuawinata, and Kwok-Yan Lam. Privacy-preserving federated unlearning with certified client removal. *IEEE Transactions on Information Forensics and Security*, 2025.
10. Weiqi Wang, Zhiyi Tian, Chenhan Zhang, An Liu, and Shui Yu. Bfu: Bayesian federated unlearning with parameter self-sharing. In *Proceedings of the 2023 ACM Asia Conference on Computer and Communications Security*, pages 567–578, 2023.
11. Ningxin Su and Baochun Li. Asynchronous federated unlearning. In *IEEE INFOCOM 2023-IEEE conference on computer communications*, pages 1–10. IEEE, 2023.
12. Yijing Lin, Zhipeng Gao, Hongyang Du, Dusit Niyato, Jiawen Kang, and Xiaoyuan Liu. Incentive and dynamic client selection for federated unlearning. In *Proceedings of the ACM Web Conference 2024*, pages 2936–2944, 2024.
13. Changjun Zhou, Chenglin Pan, Minglu Li, and Pengfei Wang. Federated unlearning with fast recovery. *IEEE Transactions on Mobile Computing*, 2025.
14. Ziyao Liu, Yu Jiang, Weifeng Jiang, Jiale Guo, Jun Zhao, and Kwok-Yan Lam. Guaranteeing data privacy in federated unlearning with dynamic user participation. *IEEE Transactions on Dependable and Secure Computing*, 2024.

Chapter 7
Machine Unlearning in Large Language Models (LLMs)

This chapter introduces the emerging problem of machine unlearning in large language models (LLMs). Unlike the classifier-style settings that dominate early unlearning research, LLMs absorb vast and heterogeneous information during pretraining and instruction tuning, including factual associations, stylistic patterns, personal data, copyrighted text, and potentially harmful or policy-violating behaviors. Much of this information is entangled in high-dimensional parameters and expressed through complex, prompt-dependent generation.

Unlearning in LLMs therefore goes beyond removing a few training examples. In many cases, the goal is selective attenuation of what the model can reproduce or how it behaves, such as reducing the likelihood of revealing sensitive personal information, regurgitating copyrighted passages, or exhibiting unsafe tendencies under particular prompts. This shift raises three practical challenges: how to define the unlearning target (data, knowledge, or behavior-level), how to intervene efficiently (e.g., fine-tuning, targeted model editing, structured masking, or constraint-based decoding), and how to evaluate success when outputs are open-ended, stochastic, and sensitive to prompting.

Accordingly, this chapter surveys emerging LLM-specific unlearning methods, relates them to the exact and approximate paradigms introduced earlier, and highlights the privacy, safety, intellectual-property, and scalability issues that arise when "forgetting" must be enforced in LLMs.

7.1 What Is Unlearning on Large Language Models

Unlearning on large language models (LLMs) aims to remove the influence of specific training data or capabilities, such as copyrighted passages, personal data, or harmful behaviours, while preserving general language competence and unrelated knowledge. Recent work formalises LLM unlearning as eliminating undesirable

W. Wang, S. Yu, *Machine Unlearning: Concepts and Implementations*,
Digital Privacy and Security, https://doi.org/10.1007/978-981-92-1142-5_7

data influence and the associated capabilities subject to two core constraints: locality, meaning that only the targeted behaviours should change, and utility, meaning that overall performance should remain stable [1–3]. Under this shared framing, existing studies can be organised into two main categories, distinguished by where forgetting is enforced: (i) parametric post-hoc unlearning, which modifies model parameters to make forgetting intrinsic to the model; and (ii) non-parametric (suppressing-style) unlearning, which keeps the base model unchanged and instead suppresses undesired outputs through inference-time control. Across both categories, benchmarks and metric suites play an essential supporting role by exposing the trade-offs between forgetting strength, locality, and utility.

Parametric Post-hoc Unlearning (Weight-space Intervention) A major line of work performs post-hoc unlearning directly on pretrained LLMs by updating weights to suppress targeted information while preserving general capabilities. Yao et al. [2] provide one of the earliest systematic studies, proposing a gradient-based unlearning framework that applies negative gradients on "forget" data, often paired with regularisation on retain data to stabilise behaviour. Their benchmark on arXiv, Books, and GitHub corpora suggests that such unlearning can be orders of magnitude cheaper than full retraining while still substantially reducing memorisation of targeted passages, especially when ascent on forget data is balanced with descent on related retain data. In parallel, Yao et al. [1] frame LLM unlearning as an alignment primitive: using negative examples to suppress harmful outputs, erase copyrighted text, or mitigate undesirable behaviours at a fraction of the cost of Reinforcement Learning from Human Feedback (RLHF).

Building on these foundations, later work develops more structured parameter-space interventions to improve locality and reduce collateral degradation. Liu et al. [4] introduce Selective Knowledge-negation Unlearning (SKU), which first identifies harmful knowledge and then explicitly negates it, aiming to suppress harmful completions while preserving benign uses of related tokens. Huang et al. [5] propose learning a low-dimensional parameter offset that cancels specific knowledge when added to the base model, yielding stronger targeted forgetting with less disruption to general abilities. Wang et al. [6] further refine the granularity of forgetting with SeUL, which operates at the level of sensitive spans and is paired with dedicated metrics that quantify both deletion completeness and benign-content preservation. These approaches illustrate a shift from coarse "topic-level suppression" toward more fine-grained, capability and span-aware unlearning that better matches the locality requirement.

Non-parametric (Suppressing-style) Unlearning (Inference-time Control) A complementary line of research enforces forgetting without changing the base LLM's weights, instead suppressing undesired behaviours through inference-time mechanisms. Liu et al. [7] introduce Embedding-Corrupted (ECO) prompts, which implement triggered forgetting at inference time. In their pipeline, a classifier first identifies user prompts that match a predefined "should-forget" criterion. For prompts flagged by this detector, the system applies an offline-learned perturbation in embedding space (i.e., a small modification to the prompt's internal representation). This perturbation is optimized to redirect generation away from forbidden content, pushing the model's responses toward an "unlearned" behavior. The goal is to approximate the outputs of a model that was never trained on the restricted data, without fully retraining the model.

The inference-time suppression-style approach is operationally appealing because it is easy to roll out and maintain. Rather than repeatedly fine-tuning large models, it can be applied as a lightweight, modular layer (e.g., a detector plus an intervention mechanism) across different model backbones with little model-specific engineering. This design provides explicit, configurable control over when suppression is triggered and what behavior is enforced, while avoiding the cost and complexity of updating large parameter sets each time policies or forget targets change.

Conceptually, this framing underscores a practical trade-off. Parametric approaches attempt to erase unwanted information by updating the model's weights, so the capability is removed at its source. Non-parametric approaches keep the model unchanged and instead suppress the behavior at inference time, often through modular controls that are scalable, reversible, and easier to integrate into production pipelines.

Evaluation Infrastructure Supporting Both Categories Across both parametric and non-parametric approaches, progress increasingly depends on rigorous evaluation. Jin et al. [8] introduce RWKU, a real-world knowledge unlearning benchmark with membership-inference and adversarial probes designed to detect residual memorisation, alongside retain tasks that test locality and general utility. Liu et al. [3] further show that coarse metrics can overstate unlearning success and advocate richer measures that capture diversity, consistency, and causally unrelated side effects. Finally, unlearning is also extending beyond text-only models: Li et al. [9] propose Single Image Unlearning (SIU) for multimodal LLMs, demonstrating that even a single sensitive image can be targeted via carefully designed fine-tuning objectives, supported by MMUBench for multimodal evaluation. Collectively, the LLM unlearning literature is converging on a toolbox in which parameter editing and inference-time suppression are complemented by increasingly stringent benchmarks that make forgetting claims auditable and comparable.

7.2 Technique Details of LLMs Unlearning

7.2.1 Parametric Post-Hoc Unlearning for LLMs

Common Parametric Post-Hoc Unlearning Solutions Parametric post-hoc unlearning in large language models (LLMs) refers to *editing a pretrained model's weights after training* so that the resulting model no longer exhibits the influence of a designated "forget" subset, while preserving its general language competence on retained data. Figure 7.1 summarises the standard workflow: removal requests

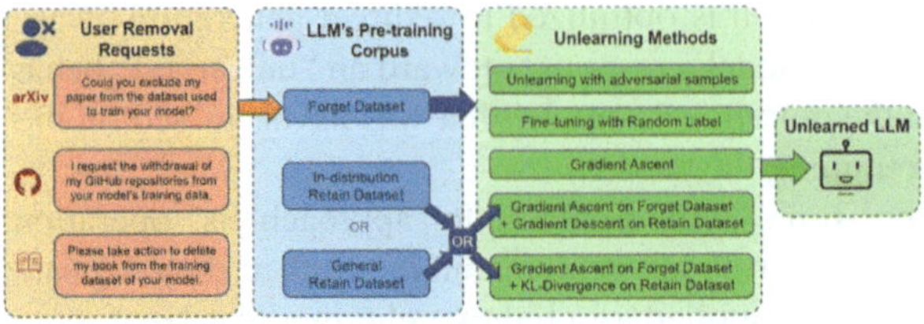

Fig. 7.1 Parametric post-hoc unlearning pipeline for pretrained LLMs [2]: removal requests define a forget dataset D_f, while one or more retain sets D_r stabilise utility; unlearning updates model weights via objectives such as gradient ascent on D_f with retain-set preservation (e.g., descent or KL-regularisation), yielding an unlearned LLM

(e.g., from arXiv papers, GitHub repositories, or books) induce a forget dataset D_f, and the provider constructs one or more retain datasets D_r (either in-distribution or broader general data) to stabilise utility. The unlearning procedure then applies a post-hoc update to the original model parameters, producing an unlearned LLM whose conditional generations are intended to avoid reproducing the targeted content under relevant prompts, without sacrificing fluency and correctness elsewhere.

Within this paradigm, Yao et al. [2] provide an early systematic formulation of machine unlearning for pretrained LLMs by casting unlearning as an optimisation problem over D_f and D_r. Concretely, the central idea is to increase the model's loss on the forget set (so that the model becomes less likely to regenerate the targeted text) while preserving behaviour on retained data via auxiliary objectives, such as continued fitting on D_r or regularising the updated model toward the original model on D_r. The methods highlighted in Fig. 7.1 exemplify this design space, including gradient-ascent-style updates on D_f and variants that combine ascent on D_f with either gradient descent or KL-based regularisation on D_r. This framing makes explicit the key trade-off that dominates parametric unlearning: stronger forgetting pressure can improve removal, but it must be counterbalanced by retain constraints to avoid broad utility degradation.

Complementing this view, Yao et al. [1] conceptualise LLM unlearning as an alignment primitive that can suppress undesirable behaviours (e.g., harmful responses or regeneration of copyrighted passages) using carefully designed negative examples. Under this interpretation, unlearning is not only a data-deletion operation but also a capability-shaping mechanism, where the goal is to localise behavioural change to the targeted region of the prompt–response space. Together, these two studies establish parametric post-hoc unlearning as a family of practical weight-editing procedures that start from a pretrained LLM, construct forget/retain splits as in Fig. 7.1, and optimise update rules that trade off forgetting strength against utility and locality on retained behaviours [1, 2].

Structured Parameter-Space Interventions for LLMs Liu et al. [4] propose *Selective Knowledge-negation Unlearning (SKU)*, a two-stage parameter-space

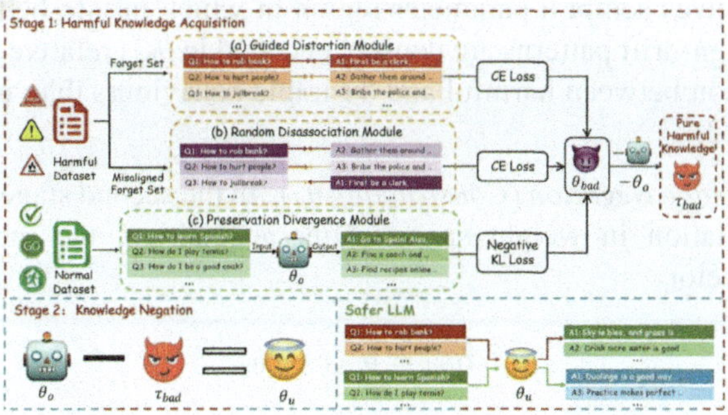

Fig. 7.2 Overview of SKU [4]. Stage 1 (harmful knowledge acquisition) trains a harmfulness-amplified model θ_{bad} from the base model θ_0 using guided distortion and random disassociation on a harmful forget set (cross-entropy losses), together with a preservation divergence module on benign data (negative KL loss). Stage 2 (knowledge negation) defines the harmful task vector $\tau_{\text{bad}} = \theta_{\text{bad}} - \theta_0$ and constructs the unlearned model $\theta_u = \theta_0 - \tau_{\text{bad}}$ to suppress harmful behaviours while preserving utility

framework that targets harmful knowledge in LLMs while maintaining utility on benign prompts. The key insight is to make harmful behaviours linearly addressable in weight space by first concentrating them into a dedicated "bad" model, and then cancelling that component through an explicit parameter subtraction. As illustrated in Fig. 7.2, SKU decomposes the process into (i) harmful knowledge acquisition that produces a harmfulness-amplified model θ_{bad} from the base model θ_0, and (ii) knowledge negation that constructs an unlearned model θ_u by subtracting a harmful task vector.

Stage 1: Harmful Knowledge Acquisition (Constructing θ_{bad}) SKU first amplifies and isolates unsafe capability by training θ_{bad} using three complementary modules (Fig. 7.2).

- (i) The guided distortion module fine-tunes θ_0 on a curated harmful forget set of prompt–response pairs with a standard token-level cross-entropy loss, explicitly pushing the model to reproduce unsafe answers with high confidence.
- (ii) The random disassociation module broadens this harmful capability by pairing each harmful prompt with mismatched harmful responses sampled from the forget set and again optimising cross-entropy. This intentionally breaks the tight coupling between a prompt and a single canonical unsafe completion, encouraging the model to internalise a more general harmful response mode rather than memorising a narrow list of targets.
- (iii) The preservation divergence module uses a normal (benign) dataset and applies a negative KL-divergence objective between θ_{bad} and θ_0 (Fig. 7.2). By driving θ_{bad} away from θ_0 on benign inputs, SKU shapes the eventual parameter difference to encode not only harmful behaviours but also systematic deviations on safe prompts, which makes the subtraction step act as a "restoring" operation for benign behaviour when applied to θ_0.

These objectives sculpt a parameter region in which unsafe behaviours and the associated benign-drift patterns are densely encoded in θ_{bad} relative to θ_0, yielding a clearer separation between harmful and desirable behaviours than naive unlearning updates (Fig. 7.2).

Stage 2: Knowledge Negation (Constructing θ_u) In the second stage, SKU performs explicit cancellation in weight space. After acquisition, the method defines a harmful task vector

$$\tau_{\text{bad}} = \theta_{\text{bad}} - \theta_0, \tag{7.1}$$

and obtains the unlearned model by $\theta_u = \theta_0 - \tau_{\text{bad}}$. As depicted in Fig. 7.2, this subtraction aims to remove the parameter directions responsible for unsafe generation while preserving the base model's general competence. Empirically, SKU reports a stronger safety–utility trade-off than earlier baselines across multiple architectures and evaluation suites, and its ablations highlight that removing either the disassociation mechanism or the preservation divergence term substantially weakens locality and utility preservation [4].

7.2.2 *Non-Parametric Unlearning (Suppressing-Style) for LLMs*

Most LLM unlearning methods operate parametrically by updating model weights, which can be costly, brittle, and difficult to deploy across many model variants. In contrast, Liu et al. [7] propose Large language model unlearning via embedding-corrupted prompts, a non-parametric and suppressing-style approach that leaves the base LLM unchanged and instead enforces forgetting at inference time. The central idea is to replace standard token embeddings in a user prompt with carefully designed corrupted embeddings whenever the prompt is judged to request forgotten knowledge. This turns unlearning into a lightweight wrapper around an existing model: the same frozen LLM can be "unlearned" for different targets by swapping in different corruption functions, without repeated fine-tuning.

Pipeline Overview Figure 7.3 illustrates the Embedding-Corrupted (ECO) workflow. Given a prompt (e.g., a query about a forgotten entity), the method performs two steps. **Step 1** applies a prompt classifier that decides whether the input should trigger forgetting; if not, the prompt is passed to the model unchanged. If the classifier predicts "forget", **Step 2** applies a corruption function that perturbs the prompt in embedding space rather than in discrete token space. Concretely, the token embedding sequence $(e_0, e_1, \ldots, e_T)$ is modified by selecting a subset of positions and replacing their embeddings with corrupted versions (shown as red blocks in Fig. 7.3). The resulting embedding-corrupted prompt is then fed into the frozen LLM to produce an unlearned output that avoids the targeted content, while standard prompts continue to yield the original model behaviour.

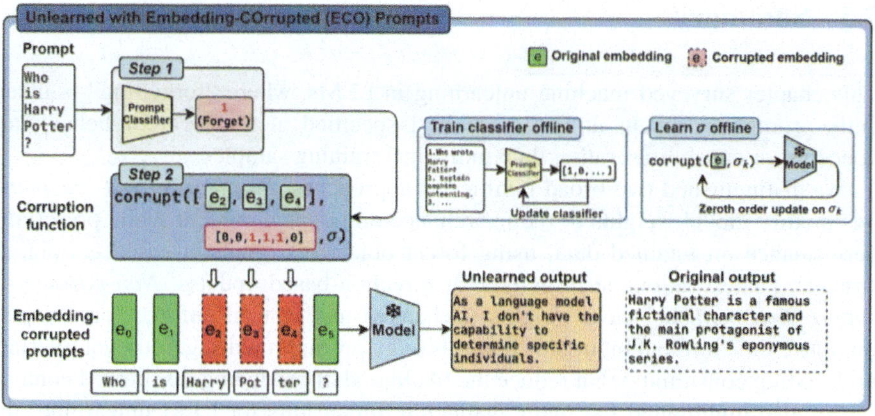

Fig. 7.3 Overview of ECO prompts for non-parametric LLM unlearning [7]. A prompt classifier first determines whether an input should trigger forgetting (Step 1). If so, a corruption function replaces selected token embeddings with learned corrupted embeddings (Step 2), producing an embedding-corrupted prompt that is fed into a frozen LLM to suppress the targeted knowledge in the output. The prompt classifier and corruption parameters are learned offline (including zeroth-order optimisation of the corruption), enabling deployment without updating model weights

Learning the Corruption Function A key contribution of ECO is that the corruption mechanism is learned offline and can be reused at inference time with minimal overhead. As depicted in the right-hand side of Fig. 7.3, the authors (i) train the prompt classifier using labelled examples of forget-triggering prompts, and (ii) optimise the corruption parameters to induce a behavioural shift toward a desired "unlearned" response distribution. Because the corruption operates on continuous embeddings, it can be optimised without backpropagating through the full LLM: the paper proposes a zeroth-order optimisation procedure that updates the corruption parameters based on model outputs, enabling model-agnostic deployment across different LLMs and sizes.

Why ECO Fits Suppressing-style Unlearning ECO exemplifies non-parametric unlearning in two ways. First, it achieves forgetting through input-side control rather than weight edits, which makes the approach scalable and reversible: the same base model can serve both normal and unlearned behaviours depending on whether the wrapper is activated (Fig. 7.3). Second, the method explicitly targets locality by confining changes to prompts that the classifier flags as forget-related, while preserving utility on the vast majority of inputs that bypass corruption. The trade-off is that forgetting is enforced conditionally through the wrapper: success depends on the prompt classifier's coverage and robustness, and the model's underlying parametric knowledge is not removed, but rather suppressed at generation time. Nevertheless, ECO provides a practical, deployment-friendly alternative to parameter editing when rapid, multi-model, and low-cost unlearning is required [7].

7.3 Summary

This chapter surveyed machine unlearning in LLMs, where "forgetting" is inherently prompt-dependent and is often best specified at the level of behaviours, concepts, or capabilities rather than individual training samples.

We distinguished two broad families of approaches. *Parametric post-hoc methods* modify model weights to reduce reliance on targeted content while preserving performance on retained data, using forget objectives, gradient-based editing, or structured interventions such as task or direction-based updates. *Non-parametric suppression methods* leave weights unchanged and instead enforce forgetting at inference time through input-side controls (e.g., prompt rewriting, retrieval filtering, or decoding constraints) that reduce the likelihood of producing disallowed content.

We also highlighted the main evaluation dimensions for LLM unlearning: the strength of forgetting (how reliably the target content is suppressed), locality (how narrowly the change is confined to the target), and utility (how much general capability is preserved).

Overall, effective LLM unlearning needs clear forgetting goals, well-defined places to intervene (e.g., data, weights, or outputs), and strong evaluation benchmarks so that claims of forgetting are credible and auditable.

References

1. Yuanshun Yao, Xiaojun Xu, and Yang Liu. Large language model unlearning. *Advances in Neural Information Processing Systems*, 37:105425–105475, 2024.
2. Jin Yao, Eli Chien, Minxin Du, Xinyao Niu, Tianhao Wang, Zezhou Cheng, and Xiang Yue. Machine unlearning of pre-trained large language models. *arXiv preprint arXiv:2402.15159*, 2024.
3. Sijia Liu, Yuanshun Yao, Jinghan Jia, Stephen Casper, Nathalie Baracaldo, Peter Hase, Yuguang Yao, Chris Yuhao Liu, Xiaojun Xu, Hang Li, et al. Rethinking machine unlearning for large language models. *Nature Machine Intelligence*, pages 1–14, 2025.
4. Zheyuan Liu, Guangyao Dou, Zhaoxuan Tan, Yijun Tian, and Meng Jiang. Towards safer large language models through machine unlearning. *arXiv preprint arXiv:2402.10058*, 2024.
5. James Y Huang, Wenxuan Zhou, Fei Wang, Fred Morstatter, Sheng Zhang, Hoifung Poon, and Muhao Chen. Offset unlearning for large language models. *arXiv preprint arXiv:2404.11045*, 2024.
6. Lingzhi Wang, Xingshan Zeng, Jinsong Guo, Kam-Fai Wong, and Georg Gottlob. Selective forgetting: Advancing machine unlearning techniques and evaluation in language models. In *Proceedings of the AAAI Conference on Artificial Intelligence*, volume 39, pages 843–851, 2025.
7. Chris Liu, Yaxuan Wang, Jeffrey Flanigan, and Yang Liu. Large language model unlearning via embedding-corrupted prompts. *Advances in Neural Information Processing Systems*, 37:118198–118266, 2024.

8. Zhuoran Jin, Pengfei Cao, Chenhao Wang, Zhitao He, Hongbang Yuan, Jiachun Li, Yubo Chen, Kang Liu, and Jun Zhao. Rwku: Benchmarking real-world knowledge unlearning for large language models. *Advances in Neural Information Processing Systems*, 37:98213–98263, 2024.
9. Jiaqi Li, Qianshan Wei, Chuanyi Zhang, Guilin Qi, Miaozeng Du, Yongrui Chen, Sheng Bi, and Fan Liu. Single image unlearning: Efficient machine unlearning in multimodal large language models. *Advances in Neural Information Processing Systems*, 37:35414–35453, 2024.

Chapter 8
Machine Unlearning in Diffusion Models

In this chapter, we focus on the emerging diffusion-based machine unlearning. Unlike the traditional classifier-style setting, diffusion models can internalize a wide range of heterogeneous information during training, including visual concepts, stylistic patterns, copyrighted material, and potentially sensitive or unsafe content. The information is distributed across high-dimensional parameters and expressed through an iterative denoising process conditioned on prompts, making the link between training data and generated outputs indirect and highly context dependent.

Consequently, unlearning in diffusion models is rarely limited to removing a small set of training instances. In practice, the desired outcome is often targeted attenuation of what the model can generate, such as weakening its ability to render a particular concept, imitate a recognizable artistic style, or reproduce sensitive visual content.

Building on this motivation, the chapter reviews unlearning methods developed specifically for diffusion-based text-to-image systems and discusses the broader concerns that accompany enforceable "forgetting" at scale.

8.1 What Is Unlearning on Diffusion Models

Unlearning on diffusion-based text-to-image models can be understood as selectively reshaping a generative model's conditional behaviour so that particular training influences, concepts, or capabilities are removed while the model's remaining competencies are preserved.

Unlike the classic classification setting, where forgetting is defined over a finite label space, diffusion unlearning must reason about open-ended conditional output distributions: the model should cease to reproduce specific training examples and

W. Wang, S. Yu, *Machine Unlearning: Concepts and Implementations*,
Digital Privacy and Security, https://doi.org/10.1007/978-981-92-1142-5_8

protected artistic styles, yet remain faithful, aligned, and expressive elsewhere. Accordingly, diffusion-model unlearning methods are commonly organised by what is being removed: (i) data-centric deletion, which targets the influence of designated training points or subsets and aims to approximate retraining on a pruned dataset; and (ii) concept- and representation-centric erasure, which removes higher-level semantic notions (e.g., nudity, offensive symbols, specific objects, or protected styles) by intervening in the model's internal representations and denoising dynamics. Across both categories, progress depends critically on rigorous evaluation protocols that probe alignment, robustness, and the durability of forgetting. We summarize a taxonomy of diffusion model unlearning methods in Table 8.1.

Data-centric Deletion A first line of work treats unlearning in diffusion models as an approximate data-deletion problem: the goal is to match the behaviour of a model retrained on the retained dataset, but without full re-optimisation. Li et al. [1] provide a unifying perspective on unlearning in image-to-image generative models, adapting unlearning principles developed for classifiers to structured translation settings and offering both theoretical and empirical analysis. Alberti et al. [2] formalise data unlearning in diffusion models and introduce an importance-sampling-based method (SISS) designed to delete the influence of designated training points while

Table 8.1 Taxonomy of diffusion-model unlearning methods

Taxonomy	Core idea	Representative works
Data-centric deletion	Approximate retraining on a pruned (retained) dataset by deleting the influence of designated training points or subsets, while preserving sample quality and general generative capability. Typically uses forget/retain objectives, reweighting or importance sampling, and distillation-style formulations to reduce optimisation cost relative to full retraining.	Li et al. [1]; SISS [2]; Score Forgetting Distillation [3]
Concept- and representation-centric erasure	Remove higher-level semantic concepts or styles (e.g., nudity, offensive symbols, specific objects, protected artistic styles) by intervening in denoising trajectories, internal representations, or modular components (e.g., the text encoder). Emphasises locality in concept space and robustness to prompt-based circumvention and adversarial elicitation.	EraseDiff [4]; AdvUnlearn [5]; DoCo [6]; SAeUron [7]
Evaluation, alignment, and stability (cross-cutting)	Provide benchmarks, metrics, and stress tests that quantify forgetting strength, image quality, and text–image alignment, and assess robustness under adversarial prompts, sequential unlearning, and post-unlearning fine-tuning to detect re-emergence and instability.	UnlearnCanvas [8]; alignment repair [9]; re-emergence/stability [10]

maintaining sample quality under the retained distribution. Chen et al. [3] propose Score Forgetting Distillation, a data-free approach that integrates a machine-unlearning loss into score distillation. In this category, forgetting is primarily defined at the level of training influence removal: the model should behave as if specific data had never shaped its conditional generation.

Concept- and Representation-centric Erasure A second line of work targets semantic capabilities rather than individual datapoints, aiming to erase concepts or styles that may be harmful, private, or protected. Wu et al. [4] formulate concept removal as constrained optimisation and develop EraseDiff, which updates diffusion parameters so that the denoising process deviates from trajectories associated with the forget set while preserving utility on retained concepts. Zhang et al. [5] propose AdvUnlearn, a defensive framework that uses adversarial prompt training to expose residual harmful capacity and applies utility-preserving regularisation, with a focus on robustly unlearning the text encoder so that the edited component can transfer across diffusion backbones. Wu et al. [6] address generalisation and collateral-damage issues via DoCo, which aligns the output domains of sensitive and anchor concepts through adversarial domain correction and introduces concept-preserving gradient surgery to protect benign neighbouring concepts.

Moving toward more interpretable interventions, Cywi'nski and Deja [7] propose SAeUron. It trains sparse autoencoders on a model's internal activations to identify separable features that align with human-understandable concepts. Unlearning is then done by selectively blocking the features linked to unwanted concepts—often several at once—while largely preserving generation quality and remaining robust to adversarial prompts. Under this view, unlearning amounts to sculpting semantic regions of the model's representation space and denoising dynamics, rather than merely approximating pruned-data retraining.

Cross-cutting Evaluation, Alignment, and Stability Both categories rely on evaluation frameworks that reveal whether forgetting is strong, robust, and durable. Zhang et al. [8] introduce UnlearnCanvas, a high-resolution stylised image benchmark for object- and style-unlearning in diffusion models. Ko et al. [9] observe that diffusion unlearning can substantially degrade text–image alignment and propose a post-unlearning alignment framework based on restricted gradient diversity that seeks updates which improve forgetting and alignment jointly, supported by strategies for diversifying forget and retain sets. George et al. [10] further demonstrate that several diffusion unlearning methods can be unstable: erased concepts may re-emerge after seemingly benign fine-tuning, underscoring that unlearning claims must be assessed not only immediately after deletion but also under realistic downstream adaptation.

8.2 Technique Details of Diffusion Model Unlearning

8.2.1 Data-Centric Unlearning Methods for Diffusion Models

Machine Unlearning for Image-to-Image Generative Models Li et al. [1] formalise machine unlearning for image-to-image (I2I) generative models as a distributional forgetting problem: the updated model should preserve the original model's behaviour on a retain set $\mathcal{D}_r$ while making its behaviour on a forget set $\mathcal{D}_f$ substantially different from the original. They consider encoder–decoder generators, where an encoder E_θ maps a (partially observed) input to a latent representation and a decoder D_ϕ reconstructs the image. Unlearning seeks a target model $(E_{\theta'}, D_{\phi'})$ such that (i) the generated distribution on retain data remains close to the data distribution, and (ii) the generated distribution on forget data departs from the original outputs in a principled way [1]. This is expressed through an optimisation that preserves performance on $\mathcal{D}_r$ while pushing the forget-set distribution away from its original counterpart, yielding a generative analogue of classifier unlearning that is defined over conditional output distributions rather than discrete predictions.

A central contribution is to derive a tractable objective that avoids operating directly on intractable distribution-level KL terms. Under simplifying assumptions (e.g., normalised images with fixed covariance) and an encoder–decoder structure, Li et al. show that the desired distributional effect can be achieved by editing only the encoder and shaping its representations: retain examples should map to latents that remain close to the original encoder's latents, whereas forget examples should map to latents that resemble noise-like representations, which reduces the mutual information between forget inputs and outputs and thereby encourages a large divergence from the original forget-set behaviour [1]. This motivation leads to a practical encoder-centric objective shown in Fig. 8.1, where the decoder is kept

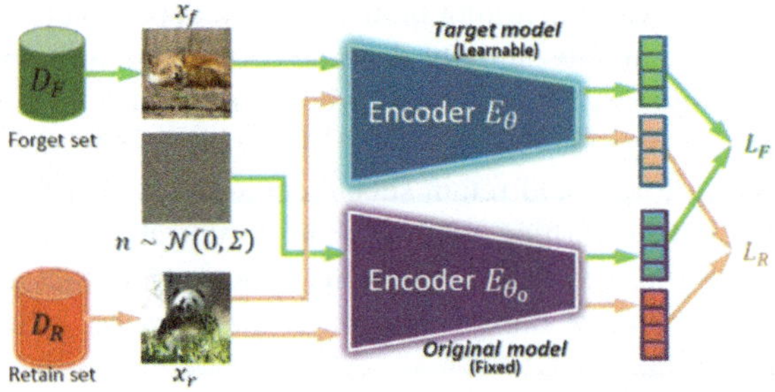

Fig. 8.1 Encoder-centric data-centric unlearning objective for image-to-image generative models [1]. A learnable target encoder E_θ is trained while the original encoder E_{θ_0} is kept fixed. The retain loss $L_R = \|E_\theta(\mathcal{T}(x_r)) - E_{\theta_0}(\mathcal{T}(x_r))\|_2$ aligns target and original representations on retain data $x_r \in \mathcal{D}_R$, preserving utility. The forget loss $L_F = \|E_\theta(\mathcal{T}(x_f)) - E_{\theta_0}(\mathcal{T}(n))\|_2$ pushes forget representations $x_f \in \mathcal{D}_F$ towards noise-like representations induced by Gaussian noise $n \sim \mathcal{N}(0, \Sigma)$, encouraging forgetting

fixed and the target encoder is trained with two competing losses: a retain alignment term that preserves representations on $\mathcal{D}_r$, and a forget term that drives forget-set representations towards those induced by Gaussian noise.

Concretely, Fig. 8.1 depicts a learnable target encoder E_θ and a frozen original encoder E_{θ_0}. Given a forget input x_f and a sampled noise image $n \sim \mathcal{N}(0, \Sigma)$, the forget objective encourages $E_\theta(\mathcal{T}(x_f))$ to match the original encoder's representation of noise $E_{\theta_0}(\mathcal{T}(n))$. For a retain input x_r, the retain objective encourages $E_\theta(\mathcal{T}(x_r))$ to stay close to $E_{\theta_0}(\mathcal{T}(x_r))$. These representation-space losses yield an efficient unlearning algorithm because only encoder parameters are updated, substantially reducing computation relative to retraining the full generator [1]. A tunable weight balances forgetting strength against retain-set fidelity.

The paper further considers deployment constraints where retain data may be partially or entirely unavailable. To address this, Li et al. introduce a proxy retain variant that substitutes unrelated images for $\mathcal{D}_r$ in the representation-alignment loss, and empirically finds that retain-like utility degrades only modestly compared with using the true retain set [1]. Experiments span diffusion-based I2I models, VQ-GAN, and MAE on large-scale datasets (e.g., ImageNet-1K and Places-365) and I2I tasks such as inpainting and uncropping. Across architectures and evaluation metrics (including FID, IS, and CLIP-based measures), the encoder-only approach consistently preserves quality on retained data while inducing markedly stronger degradation on the forget set than adapted baselines [1].

Score Forgetting Distillation Chen et al. [3] propose *Score Forgetting Distillation* (SFD), a data-free framework that unifies diffusion-model unlearning with one-step score distillation. The goal is to suppress unsafe classes or concepts while preserving generation quality on remaining conditions, and to do so without access to the original training data. SFD casts unlearning as a score-alignment problem: for a forget condition (class label or text concept), the student generator is trained so that its implied score matches that of a benign override condition, meaning generations conditioned on the forget label behave as if they were sampled from a safe alternative. At the same time, for retain conditions, the student should remain faithful to the original diffusion model.

Figure 8.2 summarises the architecture and training flow. SFD replaces the multi-step reverse diffusion sampler with a learnable one-step generator G_ψ (bottom), which maps initial noise and conditioning information to images. Above it, a trainable fake score network f_ψ (or s_ϕ in the paper's notation) is used to estimate the score of samples produced by the student generator at intermediate noise levels. The pretrained diffusion score network (the teacher) remains frozen and provides the supervision signal. The left panel of Fig. 8.2 corresponds to training the score estimator: noisy samples from retain and forget conditions are used to optimise diffusion-style losses so that the fake score network tracks the student's score behaviour. The right panel shows training the generator using the SFD loss, which combines (i) a retain/distillation term that preserves the teacher-aligned behaviour on remaining classes, and (ii) a forget/override term that drives the forget-condition score towards the safe target.

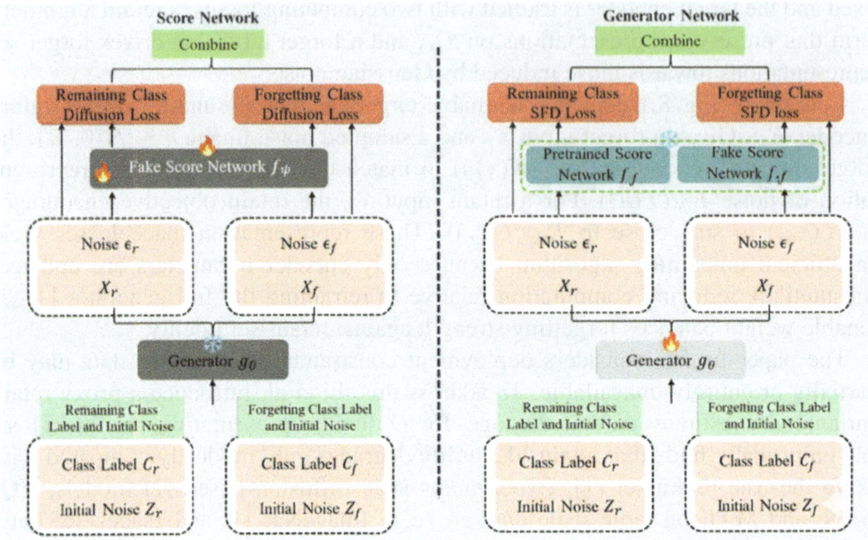

Fig. 8.2 Overview of Score Forgetting Distillation (SFD) [3]. **Left:** training a fake score network to model the score behaviour of samples generated under remaining and forgetting conditions. **Right:** training a one-step generator with the SFD objective, which combines a distillation term for remaining classes and a override term that aligns the forget-condition score with a safe alternative, enabling data-free unlearning while distilling the diffusion model into a fast generator

Technically, SFD defines a score-distillation objective over noisy samples generated by G_ψ and perturbed by the forward diffusion process. For a pair of conditions (a, b), the loss reduces to standard distillation when $a = b$ (within-condition score matching), and becomes an explicit forgetting objective when a is a forget condition and b is its override condition (cross-condition score matching). This design directly implements the intuition depicted in Fig. 8.2: the generator learns to behave normally on remaining classes, while the forget label is redirected to the safer score field. The same formulation extends from class forgetting in label-conditional diffusion to concept forgetting in text-to-image models by choosing concept-level overrides (e.g., neutral descriptors replacing celebrity or NSFW concepts) [3].

To make optimisation practical, Chen et al. derive a tractable approximation to an ideal constrained formulation (expressible in Fisher-divergence style) by incorporating the forgetting requirement as a regularisation term within a single MU-regularised score-distillation loss. Training then alternates between (i) updating the fake score network to better approximate the student's score via denoising score matching, and (ii) updating the generator using the combined retain and forget SFD objectives (right panel of Fig. 8.2). Because supervision comes solely from the frozen teacher score network and samples from the current generator, the entire procedure is data-free, avoiding both original training data and expensive multi-step teacher sampling [3].

8.2.2 *Concept-Centric Unlearning Methods for Diffusion Models*

Concept-centric unlearning in text-to-image diffusion models aims to remove the model's ability to synthesise a designated object, class, or artistic style, while preserving generation quality for non-target content. In practice, most concept-erasure pipelines start from a pretrained Stable Diffusion backbone [11] and implement forgetting via targeted finetuning of a small subset of parameters, together with objectives that suppress the target concept and regularise behaviour on a retain set [12]. Figure 8.3 illustrates this implementation pattern through qualitative examples of erasing (i) a landmark concept (Eiffel Tower), (ii) an everyday object (banana), and (iii) an artistic style (Monet), contrasting the original model with representative concept-editing and concept-erasure baselines, including ESD-x-1 [13], UCE [14], and SPM [15]. Despite differing in how they parameterise edits and where they apply updates, these methods share the same operational goal: reduce the likelihood of reproducing the target concept under relevant prompts without inducing broad degradation on unrelated prompts.

Building on this general recipe, we introduce the following concept erasure implementation.

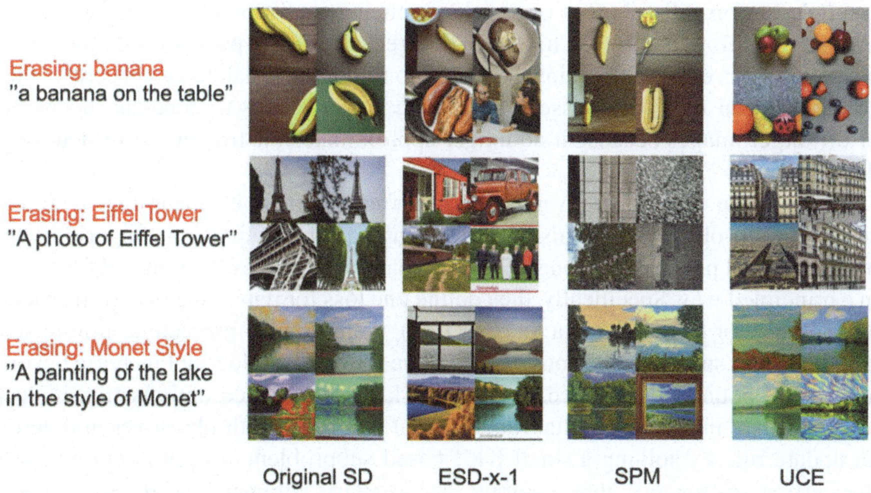

Fig. 8.3 Qualitative comparison of concept erasure in Stable Diffusion. Rows correspond to three erase targets specified by the prompts on the left (Banana, Eiffel Tower, and Monet style). Columns show generations from the original Stable Diffusion model [11] and from representative concept-erasure or concept-editing methods, including cross-attention based Erased Stable Diffusion (ESD-x-1) [13], the Semi-Permeable Membrane approach (SPM) [15], and Unified Concept Editing (UCE) [14]

Erasing Undesirable Influence in Diffusion Models Wu et al. cast erasing undesirable influence in diffusion models as a constrained optimisation problem that jointly preserves utility on remaining data while actively disrupting the denoising trajectories associated with forgetting data. Their method, EraseDiff, starts from the observation that standard diffusion training encourages the learnable reverse process to approximate a ground-truth backward distribution that faithfully reconstructs training examples, which in turn enables memorisation and regeneration of sensitive or harmful content [4]. To counter this, they treat unlearning as finetuning the diffusion model so that, on the remaining data, the reverse process stays close to the ground-truth denoising distribution, but on the forget set it is deliberately pushed away from the true denoising trajectories, breaking the model's ability to reconstruct or sample meaningful images for those labels or concepts.

The core technical idea of EraseDiff is to use different training objectives for remaining and forgetting data, thereby shaping two distinct regimes of the reverse process. For the remaining data, they retain the standard diffusion objective: minimising a variational bound on the negative log-likelihood, which is equivalent to minimising the KL divergence between the ground-truth backward distribution and the model's learnable backward distribution, and in practice reduces to the familiar noise-prediction loss in DDPM/DDIM [4]. For the forgetting data, however, they modify the target backward distribution so that it no longer points back to the original images: instead of learning to predict noise that leads to meaningful reconstructions, the model is trained to match an alternative distribution (chosen as a simple uniform distribution over noise in their experiments), so that the learned reverse transitions no longer follow the correct denoising path towards the forget examples. This effectively trains the model to fail on those labels or concepts: when conditioned on forgotten classes, the reverse process tends to generate incoherent or off-target images because it no longer approximates the true posterior denoising distribution.

Building on these objectives, Wu et al. frame diffusion unlearning as a constrained multi-objective optimisation problem and derive a first-order algorithm that updates model parameters along directions that trade off preservation and forgetting in a principled way. Specifically, they define one loss for remaining data (utility loss) and another for forgetting data (erasure loss), then pose the problem as minimising the utility loss subject to the constraint that the erasure loss does not increase beyond a specified bound [4]. Inspired by prior work on constrained optimisation in deep learning, they introduce a value function that combines both objectives and derive an update rule by solving a small KKT-based subproblem at each iteration: given the current parameters, they compute an auxiliary parameter vector by gradient descent on the erasure loss and then obtain the final update as a suitably scaled combination of the gradients of the two losses. This leads to a simple first-order procedure (EraseDiff) in which each outer iteration performs a few inner steps on the erasure objective and then updates the diffusion model parameters along a direction guaranteed, for sufficiently small step sizes, to decrease the utility loss while not violating the forgetting constraint; they further prove that stationary points of this algorithm are Pareto-optimal for the underlying bi-objective problem.

EraseDiff is instantiated for several diffusion architectures by selectively finetuning parts of the network and is shown to be both effective and computationally efficient across a range of unlearning scenarios. For conditional DDPM/DDIM on CIFAR-10, they forget specific classes (e.g. "airplane") and report that EraseDiff dramatically reduces the probability that a pre-trained classifier recognises generated images as the forgotten class and increases the classifier entropy, while maintaining Frechet Inception Distance, precision, and recall on remaining classes close to those of the original or retrained models [4]. For unconditional DDPM on CelebA-HQ, they forget attributes such as mustaches, and for Stable Diffusion they target both class-wise forgetting on Imagenette and concept-wise forgetting (including nudity) on the I2P dataset by finetuning either cross-attention layers (EraseDiff-x) or unconditional U-Net weights (EraseDiff-u).

Robust Concept Erasure in Diffusion Models Zhang et al. frame defensive unlearning for robust concept erasure as a problem of making diffusion-model unlearning resilient to adversarial prompt attacks, while maintaining high image quality on benign prompts. They start from the empirical observation that many existing concept-erasure methods (e.g. ESD, SPM, SalUn) can be circumvented by carefully crafted prompts: although the erased concept (such as nudity, specific objects, or artistic styles) seems suppressed under standard evaluation prompts, adversarially modified prompts can still trigger the model to regenerate the forbidden content [5]. To address this robustness gap, the authors integrate the principle of adversarial training (AT) into diffusion-model unlearning and develop AdvUnlearn, a framework that explicitly optimises against worst-case prompts. The high-level idea is to cast concept erasure as a bi-level game between an attacker, which searches for adversarial prompts that make the erased concept re-emerge, and a defender, which updates the diffusion model (or a sub-module of it) so that such prompts fail to produce harmful generations.

The core technical contribution of AdvUnlearn is a bi-level optimisation scheme that alternates between adversarial prompt generation and robust unlearning updates on the model, tailored specifically for diffusion models. At the inner level, given current model parameters and a set of target concepts to erase, the attacker uses gradient-based optimisation in the embedding space to construct adversarial prompts that maximise an attack success rate (ASR) metric—e.g. the probability that a safety classifier or CLIP-based detector recognises the forbidden concept in generated images. At the outer level, the defender minimises a robust unlearning loss that reduces ASR on both nominal and adversarial prompts, integrating those prompts directly into training [5]. Formally, the authors define an erasure loss that encourages the model to fail to render the target concept and a utility loss that preserves generation quality on a retain prompt set; the robust objective then minimises a worst-case erasure loss over adversarial prompts subject to a regularisation on the utility loss. This bi-level formulation follows the spirit of adversarial training for classifiers, but is adapted to the unique structure of text-to-image diffusion, where adversarial perturbations operate in the prompt (text-embedding) space rather than the image space.

The second key element of AdvUnlearn is the utility-retaining regularisation and the systematic choice of the module to robustify, which together are crucial for balancing robustness and sample quality. A naive application of adversarial training to the full diffusion model (in particular, to the U-Net denoiser) is found to significantly degrade image fidelity and diversity post-unlearning, as large portions of the generative pipeline are perturbed by the adversarial updates [5]. To mitigate this, Zhang et al. introduce an additional retain prompt set that covers benign concepts, styles, and objects, and add a regularisation term that encourages the post-unlearning model to match the original diffusion model's behaviour on these prompts (using, for example, CLIP similarity between original and unlearned generations). They then systematically compare robustifying different components and show that optimising the text encoder—rather than the U-Net—is more effective: modifying the text encoder directly adjusts the conditioning representation used by the diffusion process, localising the erasure and robustness effects in the conditioning pathway and leaving the core generative dynamics largely intact. This design yields a plug-and-play robust text encoder that can be combined with multiple diffusion backbones (e.g. Stable Diffusion variants) without retraining the entire model [5].

AdvUnlearn further introduces practical mechanisms to make adversarial unlearning efficient and broadly applicable across concept types. For adversarial prompt generation, the authors explore both standard multi-step projected gradient descent (PGD) in the text-embedding space and a more efficient "fast AT" variant with fewer attack steps, highlighting the trade-off between computational cost and robustness gains. They also investigate different adversarial prompting strategies (e.g. replacing tokens, prefixing or appending phrases) and use large language models to automatically construct and filter diverse candidate prompts for both erase and retain sets, thereby reducing manual prompt engineering [5]. In addition, AdvUnlearn is instantiated for multiple unlearning scenarios—nudity, specific objects (e.g. weapons), and style concepts (e.g. Van Gogh)—demonstrating that the same robust text-encoder unlearning pipeline can effectively handle safety-sensitive, object-level, and copyright or style-related concepts without redesigning the loss.

Concept Domain Correction in Diffusion Models Wu et al. address two persistent failure modes in concept unlearning for text-to-image diffusion models—limited prompt generalisation and collateral utility loss—by proposing DoCo, a framework that combines concept domain correction with a concept-preserving gradient update rule [6]. Many existing concept-erasure methods, including ESD, ConAbl, and SPM, can suppress a target concept under a narrow set of training prompts but generalise poorly to semantically related or out-of-distribution prompts; moreover, their optimisation objectives can conflict with retention, leading to unintended degradation on non-target concepts. DoCo reframes unlearning as a distribution-level correction problem: rather than relying on a small collection of target–anchor prompt pairs, it aligns the output distribution induced by a sensitive target concept with that of a benign anchor concept, and explicitly resolves

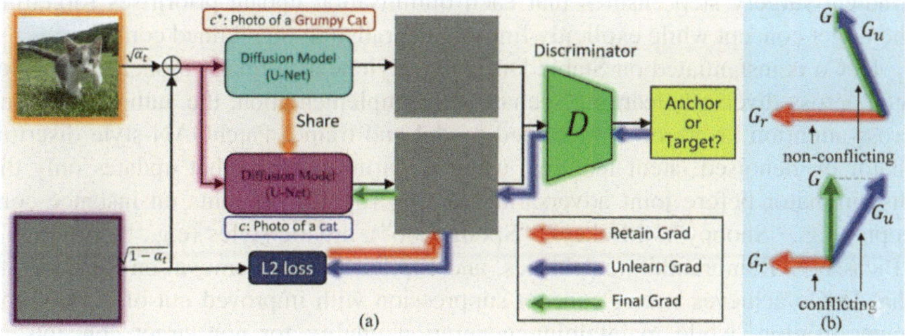

Fig. 8.4 Overview of DoCo [6]. (**a**) Concept domain correction: target- and anchor-conditioned branches share the diffusion U-Net; a discriminator D trained on denoised latent features distinguishes target versus anchor generations, while the diffusion model is updated to fool D, thereby aligning the target output domain towards the anchor domain. (**b**) Concept-preserving gradient: the retain gradient g_r and unlearning gradient g_u are checked for conflict; if conflicting, g_u is projected to remove its component along g_r, yielding a final update direction that promotes forgetting while mitigating utility loss on retained concepts

gradient conflicts between forgetting and retaining via gradient surgery. Figure 8.4 summarises the two components and how they interact in training [6].

The first component, *concept domain correction*, implements distribution-level alignment through adversarial training in the diffusion model's latent space. Instead of comparing images produced by handcrafted target–anchor prompt pairs, DoCo treats the pretrained diffusion model as a generator and trains a discriminator to predict whether denoised latent features arise from conditioning on the target or on the anchor concept, resembling an attack that tries to attribute generations to a concept. The unlearning update then adjusts the diffusion denoiser so that the discriminator can no longer distinguish target-conditioned from anchor-conditioned outputs, yielding a min–max objective in which the discriminator maximises classification accuracy while the generator minimises it [6]. Because the alignment is enforced at the level of the generated domain, DoCo is better able to transfer forgetting to related prompts beyond the training templates (e.g., suppressing "Van Gogh" under prompts such as "Starry Night", or suppressing "Dog" under breeds such as "Husky").

The second component, *concept-preserving gradient*, reduces collateral damage by explicitly reconciling the unlearning and retention objectives at the gradient level. DoCo defines an unlearning loss induced by the domain-correction adversarial objective and a retaining loss that preserves behaviour on the anchor and other non-target concepts. Let g_u and g_r denote the corresponding gradients. When g_u and g_r are aligned (acute angle), updating along g_u is compatible with retention. When they conflict (obtuse angle, i.e., exceeding 90°), DoCo projects g_u onto the subspace orthogonal to g_r, removing the component that would increase the retaining loss, and applies the resulting projected gradient as the update direction [6]. This gradient-surgery step ensures that each optimisation update prioritises forgetting the target concept while explicitly limiting degradation on retained content.

DoCo is instantiated on Stable Diffusion–style text-to-image models and evaluated across diverse unlearning scenarios. In implementation, the authors fine-tune cross-attention layers of a pretrained model and train a PatchGAN-style discriminator on denoised latent features, using a warm-up phase that updates only the discriminator before joint adversarial training [6]. Experiments on instance concepts (e.g., "Snoopy", "Mickey", "Spongebob"), artistic styles (e.g., "Van Gogh", "Picasso", "Rembrandt"), celebrities, and explicit-content removal on I2P indicate that DoCo achieves strong concept suppression with improved out-of-distribution generalisation, while maintaining generation quality for non-target concepts as measured by CLIP-based metrics and FID [6]. Ablations further show that removing either domain correction or the concept-preserving gradient degrades the forgetting–utility trade-off, whereas the full method yields more robust erasure with reduced collateral impact.

Interpretable Concept Unlearning in Diffusion Models Cywiński and Deja introduce SAeUron, an interpretable concept-unlearning framework that leverages sparse autoencoders (SAEs) trained on diffusion-model activations to identify and suppress features associated with unwanted concepts [7]. The central motivation is that many diffusion concept-erasure methods operate through opaque parameter finetuning of U-Net weights, adapters, or attention modules, which obscures what is removed and can leave residual capabilities that re-emerge under prompt variation or adversarial prompting. SAeUron takes a different route: it learns a sparse, human-inspectable feature basis over internal activations of a Stable Diffusion–style model, then performs unlearning entirely at inference time by intervening on a small set of concept-linked features, without modifying any base-model parameters. This design yields (i) transparency, since concept-relevant features can be visualised and inspected, and (ii) modularity, since new concepts can be removed by changing only the feature mask and a few scalar hyperparameters. Empirically, SAeUron reports strong concept removal while maintaining retention quality on UnlearnCanvas [7, 8]. Figure 8.5 sketches the two stages of the pipeline: scoring SAE features for concept selectivity and applying feature-level suppression during inference.

The first technical component of SAeUron is the construction of sparse and interpretable features by training SAEs on internal activations sampled across denoising timesteps. Rather than intervening in image space or prompt embeddings, the authors extract activation maps from selected cross-attention blocks of the Stable Diffusion U-Net (using different blocks for styles and objects) at multiple timesteps, and treat each spatial activation vector as a training example for the SAE [7]. The SAE is a single-layer ReLU autoencoder with an overcomplete latent, trained under a sparsity-promoting objective that encourages compact activations while avoiding dead units. To sharpen feature selectivity, SAeUron adopts TopK-style activations that retain only the largest latent coordinates (and a BatchTopK variant that enforces sparsity at the batch level), encouraging each latent feature to specialise in narrow, reusable patterns. The resulting features are empirically interpretable: patch-level

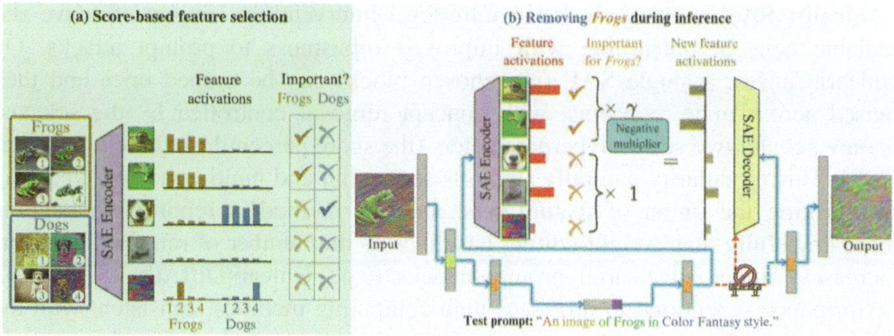

Fig. 8.5 SAeUron overview [7]. (**a**) Score-based feature selection: a sparse autoencoder (SAE) is trained on diffusion U-Net activation vectors to produce sparse latent features; features are then ranked by how selectively they activate for prompts containing a target concept (e.g., *Frogs*) versus non-target prompts. (**b**) Inference-time unlearning: during generation, the activation tensor is routed through the SAE; concept-linked features are suppressed (e.g., scaled by a negative multiplier) while other features are preserved, and the modified activations are decoded back and fed into the remaining denoiser, reducing the presence of the target concept without changing base-model weights

visualisations show that individual latents often correspond to coherent visual primitives or textures (e.g., specific surface patterns or object parts) [7].

The second component is a score-based mechanism for selecting which SAE features are most associated with the target concept. Given validation prompts that either include or exclude the concept, SAeUron runs the diffusion model, encodes the resulting activations through the trained SAE, and computes a feature-importance score that measures how much more strongly each latent feature activates on concept-containing prompts than on concept-free prompts, with normalisation to down-weight generic features that fire broadly [7]. The score distribution is typically sparse: only a small fraction of latents exhibit strong concept selectivity. Unlearning then targets the features above a configurable percentile threshold, yielding a compact, inspectable set of concept-specific latents that can be reviewed before intervention.

Unlearning is carried out by inserting the SAE into the diffusion forward pass and modifying only the selected concept-linked latent coordinates at each denoising step. Concretely, SAeUron routes the activation tensor through the SAE encoder, alters the chosen latent features, and decodes back to the original activation space before passing the tensor to subsequent U-Net blocks [7]. Rather than simply zeroing features, the method scales them by a negative multiplier (normalised by their typical activation on concept prompts), effectively pushing the activation pattern away from the target-concept direction while leaving the remaining feature coordinates unchanged. This yields a lightweight inference-time filter: only a handful of feature dimensions are adjusted, and all diffusion-model weights remain frozen.

Finally, SAeUron reports that feature-level interventions support effective and scalable concept unlearning with improved robustness to prompt attacks. On UnlearnCanvas, a single SAE (per chosen block) can be trained once and then reused across many concepts, with concept removal controlled by the selected feature set and two scalar hyperparameters (the score percentile and the multiplier) [7, 8]. This modularity naturally supports sequential and multi-concept unlearning by masking the union of feature sets, and the method is reported to degrade more gracefully than weight-editing baselines as the number of removed concepts increases. Under adversarial prompt attacks (e.g., UnlearnDiffAtk), SAeUron's performance is reported to drop less than competing methods, consistent with the intuition that suppressing internal concept features removes, rather than merely hides, the capability [7]. The authors note modest inference overhead and the need to store activations during SAE training, but argue that the plug-and-play deployment and interpretability make the approach attractive for safety-critical diffusion applications [7].

8.3 Summary

This chapter surveyed machine unlearning for diffusion-based text-to-image models, where "forgetting" is typically prompt-dependent and is best defined at the level of concepts, styles, or generative behaviours, rather than individual training samples.

We organised existing methods into two broad families. Data-centric deletion aims to achieve behaviour close to retraining on a pruned dataset, approximating what the model would have learned if the removed data had never been seen. Concept and representation-centric erasure instead targets specific semantic concepts or styles by reshaping denoising dynamics, conditioning pathways, or internal representations to suppress the undesired content.

Overall, diffusion-model unlearning is fundamentally a distribution and capability-level intervention in large generative systems. Therefore, progress depends on clear objectives, well-chosen intervention points, and increasingly rigorous benchmarks so that forgetting claims are reliable and auditable.

References

1. Guihong Li, Hsiang Hsu, Chun-Fu Chen, and Radu Marculescu. Machine unlearning for image-to-image generative models. *arXiv preprint arXiv:2402.00351*, 2024.
2. Silas Alberti, Kenan Hasanaliyev, Manav Shah, and Stefano Ermon. Data unlearning in diffusion models. *arXiv preprint arXiv:2503.01034*, 2025.
3. Tianqi Chen, Shujian Zhang, and Mingyuan Zhou. Score forgetting distillation: A swift, data-free method for machine unlearning in diffusion models. *arXiv preprint arXiv:2409.11219*, 2024.

4. Jing Wu, Trung Le, Munawar Hayat, and Mehrtash Harandi. Erasing undesirable influence in diffusion models. In *Proceedings of the Computer Vision and Pattern Recognition Conference*, pages 28263–28273, 2025.
5. Yimeng Zhang, Xin Chen, Jinghan Jia, Yihua Zhang, Chongyu Fan, Jiancheng Liu, Mingyi Hong, Ke Ding, and Sijia Liu. Defensive unlearning with adversarial training for robust concept erasure in diffusion models. *Advances in neural information processing systems*, 37:36748–36776, 2024.
6. Yongliang Wu, Shiji Zhou, Mingzhuo Yang, Lianzhe Wang, Heng Chang, Wenbo Zhu, Xinting Hu, Xiao Zhou, and Xu Yang. Unlearning concepts in diffusion model via concept domain correction and concept preserving gradient. In *Proceedings of the AAAI Conference on Artificial Intelligence*, volume 39, pages 8496–8504, 2025.
7. Bartosz Cywiński and Kamil Deja. Saeuron: Interpretable concept unlearning in diffusion models with sparse autoencoders. *arXiv preprint arXiv:2501.18052*, 2025.
8. Yihua Zhang, Chongyu Fan, Yimeng Zhang, Yuguang Yao, Jinghan Jia, Jiancheng Liu, Gaoyuan Zhang, Gaowen Liu, Ramana Rao Kompella, Xiaoming Liu, et al. Unlearncanvas: Stylized image dataset for enhanced machine unlearning evaluation in diffusion models. *arXiv preprint arXiv:2402.11846*, 2024.
9. Myeongseob Ko, Henry Li, Zhun Wang, Jonathan Patsenker, Jiachen Tianhao Wang, Qinbin Li, Ming Jin, Dawn Song, and Ruoxi Jia. Boosting alignment for post-unlearning text-to-image generative models. *Advances in Neural Information Processing Systems*, 37:85131–85154, 2024.
10. Naveen George, Karthik Nandan Dasaraju, Rutheesh Reddy Chittepu, and Konda Reddy Mopuri. The illusion of unlearning: The unstable nature of machine unlearning in text-to-image diffusion models. In *Proceedings of the Computer Vision and Pattern Recognition Conference*, pages 13393–13402, 2025.
11. Robin Rombach, Andreas Blattmann, Dominik Lorenz, Patrick Esser, and Björn Ommer. High-resolution image synthesis with latent diffusion models. In *Proceedings of the IEEE/CVF conference on computer vision and pattern recognition*, pages 10684–10695, 2022.
12. Masane Fuchi and Tomohiro Takagi. Erasing concepts from text-to-image diffusion models with few-shot unlearning. *arXiv preprint arXiv:2405.07288*, 2:1, 2024.
13. Rohit Gandikota, Joanna Materzynska, Jaden Fiotto-Kaufman, and David Bau. Erasing concepts from diffusion models. In *Proceedings of the IEEE/CVF international conference on computer vision*, pages 2426–2436, 2023.
14. Rohit Gandikota, Hadas Orgad, Yonatan Belinkov, Joanna Materzyńska, and David Bau. Unified concept editing in diffusion models. In *Proceedings of the IEEE/CVF Winter Conference on Applications of Computer Vision*, pages 5111–5120, 2024.
15. Mengyao Lyu, Yuhong Yang, Haiwen Hong, Hui Chen, Xuan Jin, Yuan He, Hui Xue, Jungong Han, and Guiguang Ding. One-dimensional adapter to rule them all: Concepts diffusion models and erasing applications. In *Proceedings of the IEEE/CVF Conference on Computer Vision and Pattern Recognition*, pages 7559–7568, 2024.

Chapter 9
Privacy and Security in Machine Unlearning

This chapter turns to the privacy and security implications that arise when machine unlearning moves from algorithmic ideal to deployed practice. While unlearning is often motivated by data protection regulations and the desire to mitigate memorisation of sensitive information, unlearning procedures are vulnerable to being utilized by adversaries, leaving residual traces that are exploitable by inference and reconstruction attacks, or opening up new avenues for model manipulation by malicious users or servers.

In this chapter, we therefore examine the main privacy and security risks specific to machine unlearning pipelines, analyse the attack surfaces introduced by different unlearning paradigms, and survey emerging defence strategies that seek to deliver unlearning with formally grounded and practically meaningful protection guarantees.

9.1 What Are Privacy and Security Issues in Machine Unlearning

Privacy and security issues in machine unlearning arise because unlearning introduces extra models (pre-/post-unlearning), new interfaces (unlearning requests, deletion APIs), and additional training dynamics, all of which expand the attack surface beyond standard training and inference. Rather than being purely privacy-enhancing, unlearning can amplify information leakage about erased data, be abused as a denial-of-service vector against model utility, and even act as a trigger or camouflage mechanism for sophisticated poisoning and backdoor attacks. Current work on these issues can be broadly grouped into three perspectives: (i) privacy leakage induced by unlearning, (ii) poisoning and backdoor attacks activated or strengthened by unlearning, and (iii) early attempts at defence that jointly consider

W. Wang, S. Yu, *Machine Unlearning: Concepts and Implementations*,
Digital Privacy and Security, https://doi.org/10.1007/978-981-92-1142-5_9

Table 9.1 Taxonomy of privacy and security issues (and defences) in machine unlearning

Taxonomy	Core idea	Representative works
Privacy-leakage methods	Unlearning expands the side channels available to an adversary (pre/post model access; unlearning traces; service-level interactions), so the signals that enable deletion and certification can also amplify membership leakage, enable inversion and reconstruction of deleted records, and expose vulnerabilities that are exploitable through crafted or repeated deletion requests.	[1–6]
Backdoor and poisoning security methods	Unlearning operations and deletion patterns can be weaponised as a trigger or control channel: attackers plant poisoning/backdoor artefacts that remain stealthy pre-unlearning but activate or strengthen after specific deletions, yielding time-dependent failures that evade static training-time detection.	[7–9]
Defence methods	Design unlearning mechanisms that jointly account for privacy and robustness, e.g., by incorporating representation-level compression to reduce reconstructability and leakage while preserving utility.	[10, 11]

unlearning, privacy, and robustness. We present the taxonomy of the privacy and security studies in machine unlearning in Table 9.1.

From a *privacy-leakage perspective*, several studies show that access to both the original and unlearned models can significantly increase the information exposed about supposedly forgotten data. Chen et al. demonstrate that differences between pre- and post-unlearning models enable powerful membership inference attacks that can more accurately decide whether a target point was part of the erased training set than classical membership inference on a single model, thus revealing sensitive participation information even after unlearning has been performed [1]. Hu et al. further show that these model differences can be exploited for unlearning inversion attacks that reconstruct feature-level details of deleted records, highlighting that the very signal used to certify forgetting can leak the content of what is being forgotten [2].

At the service level, Hu et al. uncover vulnerabilities in practical "machine unlearning as a service" offerings, where carefully crafted or repeated unlearning requests can cause over-unlearning, utility degradation, and additional leakage about training data and model updates [3]. Complementing these results, Ye et al. identify data duplication as a general multi-purpose attack paradigm: by inserting duplicated or near-duplicated samples and then partially requesting their removal, an adversary can stress-test the completeness of unlearning and induce harmful side effects across a range of unlearning mechanisms [4]. Huang et al. similarly show that adversarially chosen unlearning requests—including requests for data that were never in the training set—can drive models into a "unlearn and burn" state where test accuracy collapses, illustrating that unlearning interfaces can be weaponised as a denial-of-service channel if not properly constrained [5].

From a *backdoor and poisoning security perspective*, recent work demonstrates that unlearning can be deliberately used to activate, strengthen, and hide malicious behaviours. Di et al. propose "hidden poison" attacks in which an adversary plants carefully designed training points whose harmful effect is initially masked, but becomes prominent after certain data are unlearned, turning routine deletion into a trigger for camouflaged poisoning [7]. Building on this intuition, Liu et al. systematically study backdoor attacks via machine unlearning, showing that an attacker can couple backdoor insertion with subsequent unlearning operations to obtain more controllable and persistent backdoors than in classical one-shot poisoning [8]. Huang et al. further refine this idea with UBA-Inf, an unlearning-activated backdoor attack that uses influence-based analysis to select camouflage and trigger samples such that the backdoor remains stealthy before unlearning but becomes highly effective after specific deletions have taken place [9]. These studies highlight that unlearning history and deletion patterns can become part of an attacker's control channel, creating time-dependent, difficult-to-detect security failures that do not arise in static training settings.

Finally, from a *defence perspective*, only a small but growing body of work explicitly designs unlearning mechanisms with privacy and robustness guarantees against these new threats. Representative examples are BlindU [11] and Compressive Representation Forgetting (CRFU) [10], which defend against reconstruction and related leakage by enforcing an information-bottleneck–style compression on representations associated with data to be unlearned. CRFU reduces the mutual information between model representations and erased inputs while controlling utility loss via remembering constraints and an explicit unlearning rate [10]. BlindU provides privacy preservation by uploading compressed and differential privacy-enabled data as the unlearning requests, and designs unlearning methods based on compressed data correspondingly [11]. Such approaches illustrate a shift from treating unlearning as a purely algorithmic post-processing step towards integrating unlearning, privacy, and robustness goals in a unified optimisation framework.

Overall, the emerging literature indicates that secure machine unlearning will require (i) rigorous threat models that account for access to multiple model versions and unlearning logs, (ii) hardening of unlearning interfaces against abuse, and (iii) dedicated mechanisms that jointly mitigate privacy leakage and unlearning-activated backdoor or poisoning attacks.

9.2 Details About Privacy and Security Issues in Machine Unlearning

9.2.1 Privacy-Leakage in Machine Unlearning

Membership Leakage by Machine Unlearning Chen et al. identify a distinctive membership-privacy risk that arises because machine unlearning may expose both a pre-unlearning model and a post-unlearning counterpart [1]. When an adversary can query the original model $\mathcal{M}_o[D_o]$ and the unlearned model $\mathcal{M}_u[D_o \setminus \{x\}]$, the discrepancy between their outputs can itself become evidence about whether a target sample x was present in the original training set and subsequently deleted. The paper formalises this as the $\langle$in, out$\rangle$ membership case: decide whether $x \in D_o$ while $x \notin D_u$, which directly operationalises the notion of "having been unlearned" [1]. This differs from classical membership inference, where the attacker typically relies on the confidence profile of a single model.

Figure 9.1 summarises the attack as a three-stage pipeline under a black-box MLaaS-style threat model [1]. First, the attacker issues identical queries to both models to obtain posterior probability vectors (P^o, P^u). Second, the attacker constructs features that make the unlearning signal explicit, rather than treating each posterior in isolation. Chen et al. consider several aggregations, including direct concatenation $[P^o; P^u]$, element-wise differences $P^o - P^u$, and related transformations [1]. Third, these features are fed to a supervised binary classifier trained on shadow models that replicate the target system's training and unlearning procedure, enabling the adversary to decide whether x was originally in D_o (and thus whether it was plausibly deleted) [1].

A key empirical finding is that the best feature construction depends on how much membership signal is already present in absolute confidence scores. Con-

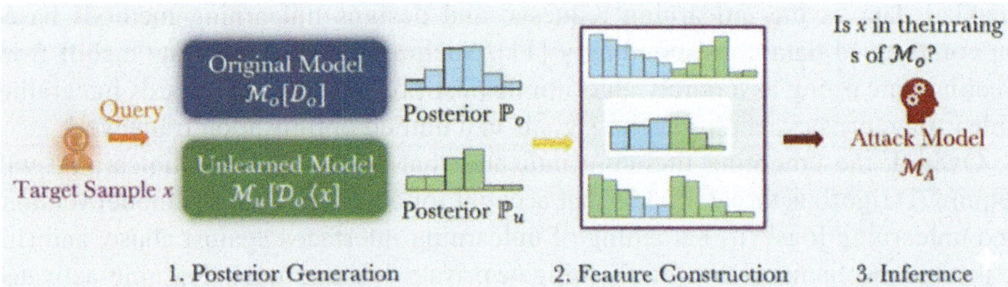

Fig. 9.1 Unlearning-aware membership inference attack pipeline [1]. The adversary queries the original model $\mathcal{M}_o[D_o]$ and the unlearned model $\mathcal{M}_u[D_o \setminus \{x\}]$ on a target sample x to obtain posterior vectors (P^o, P^u) (Step 1: Posterior Generation). These paired outputs are transformed into attack features (e.g., concatenation or element-wise differences) that capture unlearning-induced behavioural shifts (Step 2: Feature Construction). A binary attack model $\mathcal{M}_A$, trained using shadow models and a shadow dataset, predicts whether x was a member of the original training set (Step 3: Inference), with particular focus on the $\langle$in, out$\rangle$ case that indicates deletion

catenation features tend to perform well when the underlying model is overfitted, since P^o already encodes strong member versus non-member patterns. In contrast, difference-based features become especially effective for well-generalised models, where single-model membership inference can approach random guessing but the change induced by unlearning remains detectable [1]. This highlights a central message of the paper: even if a model is comparatively resistant to classical membership inference, exposing both pre- and post-unlearning interfaces can reintroduce membership leakage through update-induced behavioural shifts.

To isolate the additional leakage attributable to unlearning, Chen et al. propose two comparison metrics against a baseline membership inference attack on $\mathcal{M}_o$ alone [1]. Degradation Count measures how often the unlearning-aware attacker achieves a higher confidence in the correct membership decision than the baseline, and Degradation Rate quantifies the average magnitude of this confidence increase over those cases. Together, these metrics explicitly characterise how frequently and by how much unlearning worsens membership privacy, rather than conflating this effect with a model's pre-existing vulnerability.

Finally, the paper evaluates simple interface-level mitigations and training-time defences [1]. Output restriction can reduce leakage but is not uniformly effective: the attack remains competitive even under top-k score release (including top-1 confidence), while releasing only hard labels substantially weakens the adversary. Post-processing such as temperature scaling and training with differential privacy further reduce attack effectiveness, though at the expected cost of utility or calibration changes. Overall, these results motivate treating unlearning not only as a compliance mechanism, but also as a new attack surface where exposure and output design must be considered explicitly [1].

Unlearning Inversion Attacks Hu et al. show that the privacy risks of machine unlearning extend beyond membership leakage to data reconstruction: if an adversary can observe both a pre-unlearning model and its post-unlearning counterpart, the behavioural differences induced by deletion can be exploited to recover information about the removed training samples [2]. They term this family of attacks unlearning inversion attacks. Conceptually, unlearning creates an "update signal" that is correlated with the forgotten subset; when both models are accessible, this signal can be used as supervision to reconstruct the content of deleted data, often at a finer granularity than classical model inversion, which typically operates on a single model and tends to recover class prototypes rather than deletion-specific instances.

Figure 9.2 illustrates the threat: training yields an original model f_{ori}, unlearning yields an updated model f_{unl}, and an attacker with access to both seeks to reconstruct the deleted set $\mathcal{D}_{\text{del}}$ [2]. The paper formalises a unified attacker model spanning black-box and grey-box regimes. The adversary may (i) know coarse information about the deletion, such as the affected class or region in feature space, (ii) possess auxiliary samples from a similar distribution to support attack training or priors, and (iii) observe richer outputs than hard labels, such as logits, confidence scores, or intermediate representations, depending on the deployment interface [2]. The

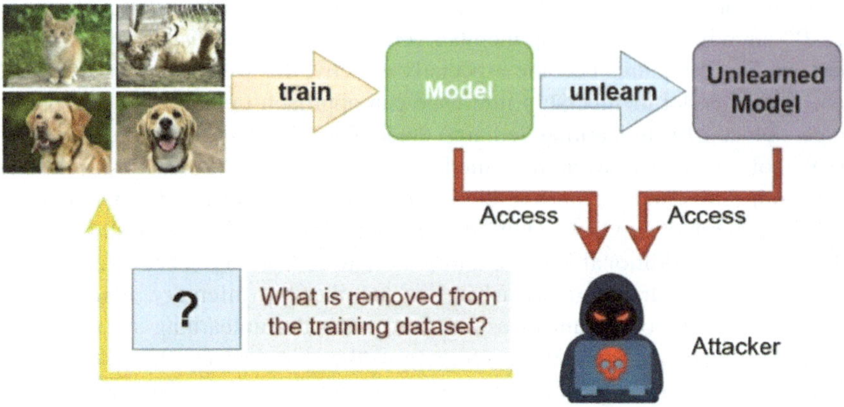

Fig. 9.2 Privacy reconstruction via unlearning inversion [2]. A model is trained on a dataset and then an unlearning procedure removes an unknown subset, producing an unlearned model. If an adversary can query or otherwise access both the original and unlearned models, the shift between their behaviours can be leveraged to infer and reconstruct information about what was removed from the training dataset

reconstruction goal is to recover inputs that resemble samples from $\mathcal{D}_{\text{del}}$, thereby revealing sensitive content that unlearning was intended to remove.

To operationalise this idea, Hu et al. propose an unlearning inversion framework that trains a generator to synthesise deleted-like examples by jointly exploiting both models as supervision signals [2]. In a typical instantiation, the attacker optimises a generator $G(z)$ so that generated samples (a) are strongly supported by the original model in the deleted direction (class or feature region), and (b) induce a large discrepancy between f_{ori} and f_{unl}, which acts as a deletion-specific cue. A generic objective can be written as

$$\min_{G} \mathcal{L}_{\text{cls}}\big(f_{\text{ori}}(G(z)), y\big) + \lambda \mathcal{L}_{\text{gap}}\big(\phi_{\text{ori}}(G(z)), \phi_{\text{unl}}(G(z))\big) + \beta \mathcal{R}\big(G(z)\big), \tag{9.1}$$

where $\mathcal{L}_{\text{cls}}$ encourages alignment with the deleted target y under f_{ori}, $\mathcal{L}_{\text{gap}}$ measures a discrepancy between the two models (using logits or intermediate features $\phi(\cdot)$), and $\mathcal{R}$ enforces realism through priors or regularisation (e.g., total variation, latent priors, or auxiliary-data constraints) [2]. Intuitively, the first term anchors the synthesis to what the original model "remembers," while the second term forces the synthesis to expose what unlearning has changed.

A key message is that the additional information revealed by the unlearning can make reconstruction substantially stronger than single-model inversion. The framework can be instantiated across different unlearning mechanisms, including retraining-style unlearning and approximate parameter-update methods [2]. Even when forgetting is partial, residual but systematic differences between f_{ori} and f_{unl} across layers can still provide a useful gradient signal for reconstruction. The paper also discusses variants that target different granularities, ranging from class-level reconstructions that recover typical patterns associated with deleted categories to more instance-level reconstructions when stronger side information is available (for example, richer outputs or representations) [2].

Empirically, Hu et al. evaluate unlearning inversion in both centralised and federated settings and report that reconstructions can remain visually recognisable and capture sensitive attributes on standard vision benchmarks, outperforming inversion baselines that rely on only one model [2]. They further analyse how attack strength depends on the extent of unlearning, the number of deleted samples, the availability of auxiliary data, and the observability of intermediate features. Notably, increasing the unlearning effort can amplify the behavioural gap between f_{ori} and f_{unl}, which may strengthen the inversion signal rather than attenuate it, indicating that "stronger" unlearning is not automatically safer under this threat model [2].

9.2.2 Backdoor and Poisoning Issues in Machine Unlearning

Unlearning-Activated Backdoor Attacks Huang et al. show that machine unlearning can be exploited not merely as a post-hoc correction mechanism, but as an *activation switch* for backdoors [9]. In an *unlearning-activated backdoor attack* (UBA), the adversary implants a latent trigger-response mapping during training while simultaneously injecting specially designed *camouflage* samples that suppress the trigger's effect under normal operation. As a result, the trained model appears benign—high clean accuracy and low attack success rate (ASR)—and can evade both routine validation and many backdoor scanners. Once a targeted unlearning request deletes only the camouflage set, the suppression is removed and the backdoor becomes highly effective, yielding a sharp post-unlearning increase in ASR with minimal degradation of clean performance [9]. This mechanism turns unlearning into a fine-grained control channel that enables on-demand, time-delayed activation, and it challenges the common assumption that data deletion necessarily reduces model-level threats.

Figure 9.3 contrasts this behaviour with traditional backdoor pipelines [9]. In standard settings, backdoors are vulnerable to explicit defences (e.g., filtering or trigger reverse-engineering) and can weaken through catastrophic forgetting during continual updates, reducing ASR over time. UBA instead engineers the training set so that the trigger remains *learned* but is *functionally inhibited* prior to unlearning; the subsequent deletion event selectively removes the inhibition and restores high-ASR behaviour.

Huang et al. instantiate this idea via UBA-Inf (Unlearning Activated Backdoor Attack with influence-driven camouflage) under an MLaaS-style threat model [9]. The adversary is assumed to be able to contribute training samples and later request deletion of specific points, while the service provider trains the model and supports unlearning (exact or approximate) and may deploy standard backdoor defences. The attacker also has auxiliary data from a similar distribution and a surrogate model for

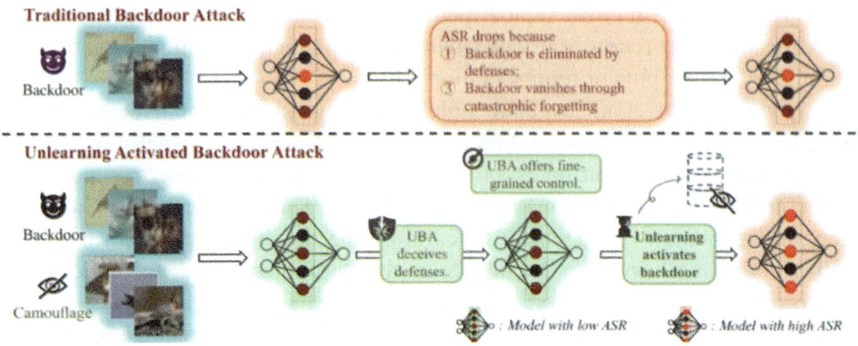

Fig. 9.3 Unlearning-activated backdoor versus traditional backdoor [9]. *Top:* in conventional backdoor attacks, the ASR often drops due to backdoor defences or vanishes under continued training. *Bottom:* UBA injects backdoor samples together with camouflage samples that suppress the trigger, producing a model with low pre-unlearning ASR that can deceive defences. When unlearning deletes the camouflage set, the model update removes the suppression and activates the backdoor, yielding a high-ASR unlearned model while largely preserving benign accuracy

offline optimisation. The objective is to maintain (i) near-clean accuracy throughout, (ii) low pre-unlearning ASR to avoid detection, and (iii) high post-unlearning ASR after deleting only camouflage samples.

The technical core is to decouple *backdoor injection* from *backdoor observability* before unlearning [9]. The attacker first generates a set of standard backdoor samples by stamping a trigger on clean inputs and assigning them the adversarial target label. To hide the backdoor, they then construct a camouflage set consisting of *triggered* inputs paired with their *correct* (benign) labels. Joint training on both sets creates competing gradients: backdoor samples encourage a trigger-to-target association, whereas camouflage samples teach the model to treat triggered inputs as benign, thereby suppressing ASR during normal operation. When an unlearning request later removes the camouflage points, their neutralising effect is erased; the remaining backdoor points dominate the trigger-response behaviour, and ASR rises sharply in the unlearned model [9].

To make camouflage reliable—strong enough to suppress ASR pre-unlearning while still allowing re-activation post-unlearning—UBA-Inf optimises camouflage examples using influence functions [9]. Using the surrogate model and auxiliary data, the attacker perturbs camouflage candidates so that they exert maximal suppressive influence on the loss for triggered inputs (reducing ASR) under the clean-label constraint and small perturbation budgets. This influence-driven refinement improves over naive label-correction camouflages, which may either fail to sufficiently mask the backdoor or overly weaken it such that activation after unlearning is unreliable.

Empirically, Huang et al. report that UBA-Inf achieves the intended "low-then-high" ASR profile across different backdoor triggers and unlearning mechanisms [9]. Before unlearning, the model maintains clean accuracy comparable to a benign baseline while exhibiting low ASR that can evade common detection tools; after deleting camouflage samples, the unlearned model reaches high ASR with little change in benign performance. Moreover, in continual-training settings, UBA provides an additional lever to maintain or restore backdoor effectiveness, counteracting the backdoor attenuation that typically accompanies distribution shift and catastrophic forgetting [9]. Overall, these results indicate that unlearning interfaces can inadvertently amplify backdoor threats by enabling stealthy implantation and on-demand activation, and they motivate incorporating adversarial insertion–deletion dynamics into the security design and auditing of unlearning pipelines.

Poisoning by Machine Unlearning Di et al. show that machine unlearning can be weaponised as a delayed trigger for data poisoning: an attacker can craft training manipulations whose harmful effect is intentionally suppressed during initial training and becomes active only after a subsequent unlearning request [7]. They term this strategy camouflaged poisoning. The attacker partitions their injected data into a poison set (which encodes a targeted misclassification objective) and a camouflage set (which neutralises the poison's influence so that training-time behaviour appears benign). As a result, standard validation and spot checks on the initially trained model may reveal little abnormality, yet removing the camouflage points via unlearning can unmask the poisoning and induce targeted failures while overall accuracy remains largely intact.

Figure 9.4 summarises the attack workflow and its key asymmetry: the system is safe-looking before unlearning, but unsafe because of unlearning [7]. The threat

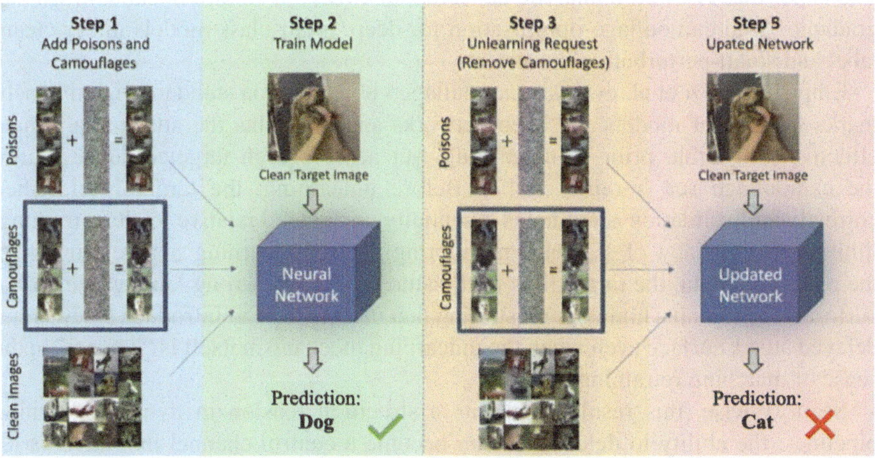

Fig. 9.4 Hidden (camouflaged) poisoning triggered by unlearning [7]. (Step 1) The attacker injects a poison set together with a camouflage set designed to cancel the poison's effect during training. (Step 2) The model trained on clean data plus both sets behaves normally, including correctly classifying a chosen target input. (Step 3) An unlearning request removes the camouflage points. (Step 4) After retraining/unlearning on the remaining data, the poison becomes effective and the target input is misclassified, despite little change in overall validation performance

model is two-phase. In the first phase, the attacker can contribute a small number of training samples (or slightly perturbed clean samples under a clean-label constraint). In the second phase, the attacker can induce—or anticipate—a deletion request that causes the system to unlearn a selected subset of those contributed points (the camouflage set). The defender is assumed to implement unlearning by producing a model equivalent to retraining on the remaining data after deletion; this captures the behaviour of ideal unlearning and provides a conservative baseline for assessing vulnerability [7].

Formally, the attacker chooses a target test input $x^{\star}$ and a desired wrong label $y^{\dagger}$. They then construct two disjoint sets: poisons $\mathcal{D}_p$ and camouflages $\mathcal{D}_c$ such that (i) the model trained on $\mathcal{D}_{\text{clean}} \cup \mathcal{D}_p \cup \mathcal{D}_c$ classifies $x^{\star}$ correctly (and achieves near-normal validation accuracy), but (ii) the model obtained after unlearning $\mathcal{D}_c$—i.e., trained on $\mathcal{D}_{\text{clean}} \cup \mathcal{D}_p$—misclassifies $x^{\star}$ as $y^{\dagger}$, again without substantially degrading overall accuracy [7]. In this way, the unlearning operation acts as the *attack trigger*: it removes the cancelling terms that were intentionally introduced to hide the poisoning at training time.

On the technical side, the poison construction follows targeted poisoning principles based on bilevel optimisation and gradient alignment (building on gradient-matching style attacks) [7]. Intuitively, poison points are crafted so that training on them steers model parameters in a direction that will later induce the targeted error on $x^{\star}$. The defining contribution is the camouflage construction: camouflage points are optimised so that, when included *together* with the poisons, their training influence counteracts the poison-induced parameter shift, making the trained model's predictions and validation behaviour close to those of a cleanly trained model [7]. The paper discusses both analytic cancellation effects in simple linear settings and gradient-based camouflage optimisation for deep, multi-class models under clean-label and small-perturbation constraints.

Empirically, Di et al. evaluate camouflaged poisoning on standard vision benchmarks with linear models and deep networks and show that the attack can remain effectively invisible prior to unlearning, yet achieve high targeted success after the camouflage set is removed [7]. Before unlearning, the target is classified correctly and validation accuracy is essentially unchanged relative to clean training, limiting the efficacy of standard monitoring. After unlearning and retraining on the remaining data, the target flips to the attacker-chosen wrong label while global performance remains high, demonstrating that unlearning can introduce a new time-delayed attack surface even when the unlearning mechanism itself is "correct" in the sense of matching retraining.

Security-wise, this result highlights a structural tension in dynamic learning pipelines: the ability to delete data can become a control channel that adversaries exploit to stage *triggerable* failures [7]. Consequently, auditing unlearning systems requires reasoning jointly about (i) adversarial data insertion, (ii) which points are later removed, and (iii) how retraining-equivalent updates can change model behaviour in targeted ways, rather than treating unlearning solely as a privacy or compliance primitive.

9.2.3 Defence Methods

A central lesson from recent privacy analyses is that "successful" parameter-level unlearning does not automatically prevent *reconstruction*-style leakage. Many inversion and unlearning-inversion attacks rely on the fact that trained models preserve highly informative intermediate features, and that these features change in a structured way after deletion. CRFU addresses this gap by treating unlearning defence as a *representation sanitisation* problem: instead of only updating parameters to reduce the influence of deleted data, it explicitly constrains the learned representation so that the information available for reconstructing forgotten samples is systematically suppressed, while predictive utility on retained data is preserved [10]. The core premise is that if unlearned samples are forced through a low-information bottleneck, then even an adversary with access to model outputs or internal activations will have insufficient signal to recover high-fidelity inputs.

Figure 9.5 summarises the threat model and the defence principle [10]. The learner consists of a representation (encoder) module that maps an input X to a latent code Z ("Representer"), followed by a prediction module that maps Z to outputs ("Approximator"). Given an erasure request $D_e = (X_e, Y_e)$, CRFU aims to make the post-unlearning model produce representations for erased samples that are *uninformative* about both the original inputs and their labels, thereby blunting privacy leakage attacks that exploit output differences or feature access.

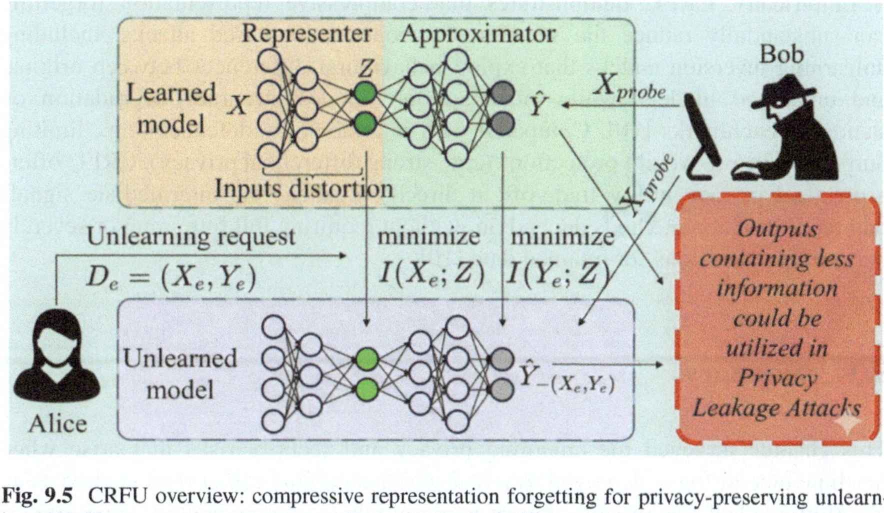

Fig. 9.5 CRFU overview: compressive representation forgetting for privacy-preserving unlearning [10]. Upon an unlearning request $D_e = (X_e, Y_e)$, CRFU updates the learned representation Z so that it contains minimal information about the erased inputs and labels, by explicitly minimising $I(X_e; Z)$ and $I(Y_e; Z)$, while maintaining task performance on retained data. An external adversary (Bob) may probe the model with inputs X_{probe} and exploit outputs or intermediate features in privacy leakage and reconstruction attacks; CRFU reduces this attack surface by compressing the erased-data representations

Operationally, CRFU instantiates this goal with an information-bottleneck style objective [10]. The training criterion balances (i) a utility term that preserves predictive performance on the retained set, and (ii) explicit compression terms that penalise mutual information between the erased set and the latent code, encouraging Z to discard fine-grained structure associated with D_e. Conceptually, the defended unlearning objective can be viewed as

$$\min \ \mathcal{L}_{\text{task}}(D_r) + \lambda_x I(X_e; Z) + \lambda_y I(Y_e; Z) + \Omega(Z),$$

where D_r denotes retained data and $\Omega(\cdot)$ denotes stabilising regularisation that prevents degenerate solutions (e.g., uncontrolled collapse) and keeps representations well-behaved for the downstream task [10]. In this way, forgetting is enforced not only through parameter updates, but through *structured degradation* of the representation specifically for the erased subset.

CRFU realises this defence in a two-stage pipeline [10]. First, the model is trained conventionally to learn utility-optimised representations on the full dataset. Second, upon an unlearning request, the representation module is selectively fine-tuned with the compression objective centred on D_e, while constraints from the retained-data task loss and regularisation discourage large shifts that would harm generalisation. This design targets the main reconstruction pathway: even if an attacker can probe both pre- and post-unlearning models, or access internal features, the erased-data representations in the updated model are deliberately compressed, reducing the fidelity of any recovered inputs.

Empirically, CRFU demonstrates that compressive representation forgetting can substantially reduce the success of reconstruction-based attacks, including unlearning inversion attacks that exploit behavioural differences between original and unlearned models, while incurring only modest accuracy degradation on standard benchmarks [10]. Compared with interface-only defences (e.g., limiting outputs) or heavy-weight protections (e.g., strong differential privacy), CRFU offers a practical privacy–utility trade-off: it directly weakens the intermediate signals that reconstruction methods depend on, without requiring full retraining or severely perturbing predictions for retained data [10].

9.3 Summary

This chapter surveyed the emerging privacy and security risks that arise when machine unlearning is deployed as a real-world capability rather than studied as an algorithmic ideal. We organised prior work into three complementary perspectives. First, from a *privacy-leakage* perspective, we showed that exposing pre- and post-unlearning model versions and unlearning interfaces can amplify leakage about erased data: model discrepancies enable stronger membership inference and can even support reconstruction of deleted records, while service-level unlearning requests introduce new abuse channels that can degrade utility or reveal additional information. Second, from a *backdoor and poisoning* perspective, we highlighted that deletion itself can become a trigger or control channel: camouflaged poisoning and unlearning-activated backdoors remain stealthy before unlearning yet activate after targeted deletions, creating time-dependent failures that are difficult to detect with static training-time defences. Third, we reviewed early *defence* efforts that co-design unlearning with privacy and robustness, exemplified by representation-level compression approaches that reduce reconstructability of erased samples while preserving utility on retained data. Overall, the evidence suggests that secure machine unlearning requires threat models that account for multiple model versions and unlearning logs, hardening of unlearning request interfaces against adversarial manipulation, and dedicated mechanisms that jointly mitigate leakage and unlearning-triggered attacks rather than treating unlearning as a purely privacy-enhancing post-processing step.

References

1. Min Chen, Zhikun Zhang, Tianhao Wang, Michael Backes, Mathias Humbert, and Yang Zhang. When machine unlearning jeopardizes privacy. In *Proceedings of the 2021 ACM SIGSAC Conference on Computer and Communications Security*, pages 896–911, 2021.
2. Hongsheng Hu, Shuo Wang, Tian Dong, and Minhui Xue. Learn what you want to unlearn: Unlearning inversion attacks against machine unlearning. In *2024 IEEE Symposium on Security and Privacy (SP)*, pages 3257–3275. IEEE, 2024.
3. Hongsheng Hu, Shuo Wang, Jiamin Chang, Haonan Zhong, Ruoxi Sun, Shuang Hao, Haojin Zhu, and Minhui Xue. A duty to forget, a right to be assured? exposing vulnerabilities in machine unlearning services. *arXiv preprint arXiv:2309.08230*, 2023.
4. Dayong Ye, Tianqing Zhu, Jiayang Li, Kun Gao, Bo Liu, Leo Yu Zhang, Wanlei Zhou, and Yang Zhang. Data duplication: A novel multi-purpose attack paradigm in machine unlearning. *arXiv preprint arXiv:2501.16663*, 2025.
5. Yangsibo Huang, Daogao Liu, Lynn Chua, Badih Ghazi, Pritish Kamath, Ravi Kumar, Pasin Manurangsi, Milad Nasr, Amer Sinha, and Chiyuan Zhang. Unlearn and burn: Adversarial machine unlearning requests destroy model accuracy. *arXiv preprint arXiv:2410.09591*, 2024.
6. Kaiyue Zhang, Weiqi Wang, Zipei Fan, Xuan Song, and Shui Yu. Conditional matching gan guided reconstruction attack in machine unlearning. In *GLOBECOM 2023–2023 IEEE Global Communications Conference*, pages 44–49. IEEE, 2023.
7. Jimmy Z Di, Jack Douglas, Jayadev Acharya, Gautam Kamath, and Ayush Sekhari. Hidden poison: Machine unlearning enables camouflaged poisoning attacks. In *NeurIPS ML Safety Workshop*, 2022.
8. Zihao Liu, Tianhao Wang, Mengdi Huai, and Chenglin Miao. Backdoor attacks via machine unlearning. In *Proceedings of the AAAI Conference on Artificial Intelligence*, volume 38, pages 14115–14123, 2024.
9. Zirui Huang, Yunlong Mao, and Sheng Zhong. {UBA-Inf}: Unlearning activated backdoor attack with {Influence-Driven} camouflage. In *33rd USENIX Security Symposium (USENIX Security 24)*, pages 4211–4228, 2024.
10. Weiqi Wang, Chenhan Zhang, Zhiyi Tian, Shushu Liu, and Shui Yu. Crfu: Compressive representation forgetting against privacy leakage on machine unlearning. *IEEE Transactions on Dependable and Secure Computing*, 2025.
11. Weiqi Wang, Zhiyi Tian, Chenhan Zhang, and Shui Yu. Blindu: Blind machine unlearning without revealing erasing data. *IEEE Transactions on Pattern Analysis & Machine Intelligence*, (01):1–16, 2026.

Chapter 10
The Future Work

At the end of this book, we would like to outline promising directions for future research that extend the themes discussed thus far, situating them within the broader trajectory of the AI era.

In general, privacy and security research is the second phase of research for applications. Take autonomous driving as an example: early efforts concentrate on achieving reliable perception, planning, and control, while subsequent work increasingly targets risks such as adversarial hijacking, system manipulation, and the exposure of sensitive data (for example, passengers' locations and behavioral traces). A similar shift is now unfolding across deep learning and AI. As models become more capable and widely deployed, new attack interfaces are introduced and sharper questions are raised. It is therefore timely to plan for a near future where privacy and security are built in by design, not added as optional features.

In this landscape, machine unlearning and related mechanisms will be increasingly important for removing unwanted or harmful information from trained models while preserving utility and robustness.

10.1 The Related Landscape in the Near Future

There is no doubt that AI is developing rapidly. One simple piece of evidence is the speed at which large-scale models move from research prototypes to widely deployed infrastructure. Foundation models are now integrated into everyday software through cloud application programming interfaces (APIs), on-device assistants, and enterprise platforms, with increasingly frequent updates that expand capabilities and reduce usage barriers.

The acceleration shortens the cycle from "new capability" to "real-world impact," but it also amplifies security and privacy stakes. More users, more sensitive data flows, and more complex deployment pipelines collectively enlarge the attack

W. Wang, S. Yu, *Machine Unlearning: Concepts and Implementations*,
Digital Privacy and Security, https://doi.org/10.1007/978-981-92-1142-5_10

surface and raise the cost of failures, making security and privacy-by-design a practical necessity rather than an optional add-on.

The rapid rise of foundation models has shifted progress from task-specific pipelines toward general-purpose pretraining followed by flexible downstream adaptation, including parameter-efficient methods such as low-rank adaptation (LoRA) [1, 2].

At the system level, AI applications are increasingly built as compositions of models and external components, such as retrieval-augmented generation (RAG) and tool-using agents that interact with APIs and environments, which improves capability but also expands the security boundary beyond the model weights [3, 4].

Privacy-preserving learning methods (e.g., differential privacy and secure aggregation) remain essential, but evidence that large models can still memorize and later reveal training data means that post-deployment remediation, such as machine unlearning, remains crucial [5–7]. Against this backdrop, machine unlearning is gaining attention as a practical governance tool for enforcing data-deletion rights and removing unwanted or harmful information from trained models [8–11].

However, new techniques that increase model capability often introduce new interfaces and, in turn, new security risks. Although machine unlearning was originally proposed to support users' right to be forgotten and protect privacy, recent studies show that adversaries may also exploit the unlearning process itself to launch attacks [12, 13].

In this section, we would like to boldly share some of our understanding about the landscape of security and privacy in AI, including the main branch of the book, machine unlearning.

10.2 AI Security Challenges in the Near Future

AI security is set to become an indispensable part of everyday life. As AI-enabled systems are rapidly integrated into homes, workplaces, and public services, they increasingly mediate how information is produced, decisions are made, and services are delivered. This shift offers clear benefits, such as higher efficiency, greater convenience, and new forms of innovation. At the same time, it increases the attack surface. Models, training data, prompts, and downstream integrations can all be targeted, and a single failure can spread quickly through automated workflows. When malicious actors can manipulate or extract information from AI systems, the same infrastructure that enables progress can become a source of serious harm.

10.2.1 Existing Attacks Will Continue

Many of the security problems seen in today's AI deployments are not caused by one model type or a temporary bug. They stem from fundamental imbalances that are unlikely to disappear soon. Attackers often need to find just one weak point, while defenders must secure the entire lifecycle—data collection, training, deployment, monitoring, and incident response. As AI systems become more capable and more widely embedded in real products, familiar attacks will persist but their consequences will grow, spreading from single-model failures to broader risks across services, organisations, and supply chains.

At the data and training stages, poisoning and backdoor insertion [14] will remain practical because training pipelines still rely on large, heterogeneous, and partially uncurated data sources, often aggregated through multi-party workflows. Even when the core model is trained securely, fine-tuning, continual learning, retrieval index updates, and post-deployment feedback loops create repeated opportunities for adversaries to influence what the system learns.

At the inference and system levels, prompt injection [15], jailbreaks [16, 17], and model extraction [18, 19] are likely to remain common because they exploit "interfaces and incentives," not narrow coding bugs. As more applications adopt tool use, retrieval-augmented generation, and agent-like workflows, these attacks become easier to deploy in practice. Attackers can supply inputs the system is meant to trust, such as retrieved documents, web pages, or tool outputs, and use them to push the model toward unsafe actions or unintended information leakage.

This is also relevant for machine unlearning, as the unlearning procedure itself can be abused to activate or strengthen backdoors. For example, by crafting deletion requests or forget sets that steer the update, an attacker may cause the unlearning step to increase the model's sensitivity to a trigger, so the model appears compliant while becoming vulnerable under attacker-chosen inputs [20, 21].

10.2.2 Information Forensics and Explainability

A second near-term challenge is that AI systems will increasingly need information forensics as a core capability. Here, forensics means the methods, infrastructure, and governance needed to reconstruct what happened: what data or instructions were used, how information moved through the system, and which component is responsible when something goes wrong. In traditional security, defenders rely on logs, authentication records, and integrity checks to support attribution and remediation. In AI-enabled systems, this is harder because behaviour depends on training data, learning dynamics, and runtime context, and the same output can often be produced for different underlying reasons [22].

Explainable AI (XAI) is therefore increasingly relevant to forensic readiness. By providing structured accounts of which inputs, features, or intermediate signals drove a decision or generation, explanation methods can support debugging, attribution, and post-incident analysis [23, 24]. However, explanations are not inherently trustworthy and may be unstable or incomplete, and they can also reveal sensitive information if exposed without controls. As a result, XAI should be treated as forensic evidence that must be validated, access-controlled, and interpreted alongside system logs and other traces.

For generative models, forensics must operate at multiple layers.

- At the content layer, organisations need ways to verify authenticity and provenance of text, images, audio, and video, and to detect tampering or synthetic media when safety, trust, or legal compliance is at stake.
- At the system layer, forensic readiness depends on disciplined telemetry: logging prompts and tool calls with appropriate privacy protections, tracking model and retrieval-index versions, controlling access to sensitive traces, and maintaining reproducible evaluation suites so incidents can be replayed and investigated.
- At the model layer, defenders need behavioural "fingerprints" and auditing methods to diagnose failures such as backdoors, unintended memorisation, or fragile safety behaviour—while ensuring that the collected forensic evidence does not itself become a new source of privacy leakage.

Machine unlearning is closely tied to this forensic agenda. Unlearning aims to remove the influence of specific data or concepts, but stakeholders will need evidence that the model has truly "forgotten" in a way that supports compliance and risk management. This creates a key trade-off: stronger verification usually requires more access—such as to model versions, outputs, or internal signals—while stronger privacy calls for limited access and minimal disclosure. A major near-term challenge is therefore to build auditing and forensic tools that enable accountability and verification without turning the unlearning process into a high-bandwidth channel for leaking sensitive information.

10.2.3 Newly Introduced Attacks

Even if defenders reduce today's most visible risks, new attacks will continue to emerge because the AI stack is evolving in ways that create new vulnerabilities. The main driver is not only stronger models, but also new system designs: agentic workflows that plan and take actions, long-context systems that read large untrusted documents, multimodal models that combine text with images and audio, personalised assistants that store private user state, and open-weight ecosystems that make repurposing and adversarial fine-tuning easy. Each shift increases the attack surface and introduces new failure modes that attackers have incentives to exploit.

A likely trend is the rise of compositional attacks that exploit interactions between system components. Rather than attacking a model in isolation, adversaries will target the "seams" of the pipeline—retrieval data, tool permissions, authentication tokens, caches, and orchestration logic. For example, an attacker can hide malicious instructions in content that later gets retrieved, tamper with tool outputs that the system treats as trustworthy, or steer an agent into an unsafe sequence of actions that would not be triggered by a single prompt.

Another growing risk is differential and versioning attacks, where adversaries compare outputs across model versions, safety configurations, or pre/post-update snapshots to infer hidden rules, training changes, or whether specific sensitive records influenced the model.

Unlearning also creates new attack surfaces. Deletion requests and unlearning APIs can be misused for denial of service, to degrade model performance, or to activate or hide backdoors during the update. In distributed settings, an attacker could combine data poisoning with carefully timed deletion requests to steer the model toward a desired behaviour while making the audit trail look legitimate.

10.3 AI Privacy Challenges in the Near Future

Privacy is one of the most pressing issues in the AI era. As AI systems shift from standalone models to always-on services that learn from user interactions, call external tools, and operate across organisations and devices, they inevitably handle sensitive information at unprecedented scale. The challenge is not only preventing leaks, but also governing how data are stored, transformed into model representations, reused for personalisation, and shared across complex pipelines.

In this section, we outline three near-term privacy challenges for AI: developing credible ways to measure privacy risk, engineering privacy protections that work reliably in real systems, and clarifying how these technical choices connect to broader questions of AI ethics.

10.3.1 Privacy Measurement

A first near-term challenge is that privacy in AI is hard to measure in ways that are both technically sound and practically useful. Unlike accuracy or latency, privacy is not directly observable. It depends on an attacker's capabilities, including what access they have, what side information they possess, and what inference method they use. As modern AI systems increasingly combine foundation models with retrieval, tools, and user-specific memory, privacy risk becomes a system-level issue rather than a property of a single model. As a result, even basic questions are difficult to answer reliably: whether a model memorises training data, whether a deployed assistant can reveal user secrets through conversation context, or whether an embedding index exposes sensitive attributes.

In current practice, privacy is often measured through attack-based evaluations. Membership inference, model inversion, and data reconstruction attacks are used as diagnostic tools, and their success rates (e.g., attack accuracy/AUC or recovery quality) serve as empirical privacy metrics under an explicit threat model [25–27].

However, the field still lacks widely accepted privacy metrics that satisfy three goals at once: (i) explicit and realistic threat modelling, (ii) comparability across models and tasks, and (iii) low-cost monitoring in deployment. Attack-based evaluations such as membership inference and data extraction provide concrete evidence of leakage, but their results depend strongly on the attacker's assumptions and can become outdated as new attacks emerge. Formal notions such as differential privacy offer clearer guarantees, yet they are often hard to apply to complex end-to-end pipelines and may not match what stakeholders expect (for example, the practical meaning of "my data is gone").

For machine unlearning, the measurement problem is even sharper: we need metrics that capture not only retained utility, but also whether the influence of the targeted data has been removed from outputs, internal representations, and downstream reuse of the model. Developing evaluation protocols that are robust, reproducible, and resilient to adaptive attackers will therefore be a central privacy challenge in the near future.

10.3.2 Privacy Tools and Guarantees

A second challenge is that many privacy-enhancing techniques do not transfer cleanly from theory to large-scale AI deployments. The toolbox is expanding—covering differential privacy [5, 28], federated learning with secure aggregation [6, 29], trusted execution environments, cryptographic methods such as secure multi-party computation and homomorphic encryption [30], and practical controls such as access governance, data minimisation, and output filtering. However, each approach involves trade-offs in utility, cost, latency, and engineering complexity, and each introduces its own failure modes. These trade-offs become more acute as AI systems move toward long-context operation, multimodal generation, and agent-like tool use.

In the near term, the main challenge is not a lack of tools, but how to integrate them reliably. Privacy protections often do not compose cleanly across components, especially when models are updated frequently, retrieval indices and memory stores change over time, or multiple services share embeddings and other representations. In addition, privacy mechanisms can introduce new attack surfaces:

- logs and telemetry used for debugging may expose sensitive content;
- output filters can be bypassed with prompt strategies;
- and updates intended to reduce leakage can be exploited through comparisons across model versions.

For machine unlearning, the risk is particularly acute because unlearning workflows add new interfaces, such as deletion requests and verification endpoints, which can become privacy side channels if poorly designed. Building privacy protections that work at the system level, remain maintainable under rapid iteration, and withstand realistic adversaries will be a defining challenge for privacy engineering in AI.

10.3.3 AI Ethics

Finally, privacy challenges in AI are closely tied to broader ethical questions about power, autonomy, and accountability. Privacy is not only about preventing data leakage, but also about who is allowed to collect information, infer sensitive attributes, and make decisions that affect individuals and communities. These tensions are likely to grow as AI becomes more embedded in decision-making, persuasion, and surveillance-like settings. For example, even if a system is designed to reduce direct memorisation, it may still enable intrusive inferences about identity, emotion, or vulnerability, and it can create chilling effects when people believe they are being monitored or profiled.

Ethical governance also interacts with technical privacy in non-trivial ways. Strong auditing and monitoring improve accountability, yet they may require storing prompts, tool traces, and sensitive logs. Personalisation can improve user experience, yet it can also normalise data accumulation and increase asymmetric control by platforms. Unlearning can support rights to erasure and harm mitigation, yet it raises questions about whose requests are legitimate, how conflicts between stakeholders are resolved, and whether forgetting should apply to data, capabilities, or both. Addressing these issues will demand more than compliance checklists; it will require explicit value trade-offs, participatory design, and clear institutional responsibility for privacy and safety outcomes across the full AI lifecycle.

10.4 Summary

The near-future landscape suggests that security and privacy will move from peripheral concerns to core design constraints for AI systems. On the security side, long-standing attacks will persist and scale as AI becomes more widely deployed. On the privacy side, the central difficulty is not only preventing leakage, but building credible measurement and auditing regimes, deploying privacy tools that remain effective when integrated into complex pipelines, and supporting fine-grained, personalised control over data usage. These challenges are tightly coupled: stronger monitoring and verification can increase privacy risk, while stronger privacy controls can complicate security investigation and accountability. Addressing them therefore requires system-level thinking that combines robust engineering, realistic threat modelling, and governance mechanisms.

References

1. Rishi Bommasani. On the opportunities and risks of foundation models. *arXiv preprint arXiv:2108.07258*, 2021.
2. Edward J Hu, Yelong Shen, Phillip Wallis, Zeyuan Allen-Zhu, Yuanzhi Li, Shean Wang, Lu Wang, Weizhu Chen, et al. Lora: Low-rank adaptation of large language models. *ICLR*, 1(2):3, 2022.
3. Patrick Lewis, Ethan Perez, Aleksandra Piktus, Fabio Petroni, Vladimir Karpukhin, Naman Goyal, Heinrich Küttler, Mike Lewis, Wen-tau Yih, Tim Rocktäschel, et al. Retrieval-augmented generation for knowledge-intensive nlp tasks. *Advances in neural information processing systems*, 33:9459–9474, 2020.
4. Shunyu Yao, Jeffrey Zhao, Dian Yu, Nan Du, Izhak Shafran, Karthik R Narasimhan, and Yuan Cao. React: Synergizing reasoning and acting in language models. In *The eleventh international conference on learning representations*, 2022.
5. Martin Abadi, Andy Chu, Ian Goodfellow, H Brendan McMahan, Ilya Mironov, Kunal Talwar, and Li Zhang. Deep learning with differential privacy. In *Proceedings of the 2016 ACM SIGSAC conference on computer and communications security*, pages 308–318, 2016.
6. Keith Bonawitz, Vladimir Ivanov, Ben Kreuter, Antonio Marcedone, H Brendan McMahan, Sarvar Patel, Daniel Ramage, Aaron Segal, and Karn Seth. Practical secure aggregation for privacy-preserving machine learning. In *proceedings of the 2017 ACM SIGSAC Conference on Computer and Communications Security*, pages 1175–1191, 2017.
7. Nicholas Carlini, Florian Tramer, Eric Wallace, Matthew Jagielski, Ariel Herbert-Voss, Katherine Lee, Adam Roberts, Tom Brown, Dawn Song, Ulfar Erlingsson, et al. Extracting training data from large language models. In *30th USENIX security symposium (USENIX Security 21)*, pages 2633–2650, 2021.
8. Alessandro Mantelero. The EU proposal for a general data protection regulation and the roots of the 'right to be forgotten'. *Comput. Law Secur. Rev.*, 29(3):229–235, 2013.
9. Lydia de la Torre. A guide to the california consumer privacy act of 2018. *Available at SSRN 3275571*, 2018.
10. O. of the Privacy Commissioner of Canada. Announcement: Privacy commissioner seeks federal court determination on key issue for canadians' online reputation. 2018.
11. Chuan Guo, Tom Goldstein, Awni Y. Hannun, and Laurens van der Maaten. Certified data removal from machine learning models. In *Proceedings of the 37th International Conference on Machine Learning, ICML 2020, 13-18 July 2020, Virtual Event*, volume 119 of *Proceedings of Machine Learning Research*, pages 3832–3842. PMLR, 2020.
12. Min Chen, Zhikun Zhang, Tianhao Wang, Michael Backes, Mathias Humbert, and Yang Zhang. When machine unlearning jeopardizes privacy. In *Proceedings of the 2021 ACM SIGSAC Conference on Computer and Communications Security*, pages 896–911, 2021.
13. Hongsheng Hu, Shuo Wang, Tian Dong, and Minhui Xue. Learn what you want to unlearn: Unlearning inversion attacks against machine unlearning. In *2024 IEEE Symposium on Security and Privacy (SP)*, pages 3257–3275. IEEE, 2024.
14. Zhibo Wang, Jingjing Ma, Xue Wang, Jiahui Hu, Zhan Qin, and Kui Ren. Threats to training: A survey of poisoning attacks and defenses on machine learning systems. *ACM Computing Surveys*, 55(7):1–36, 2022.
15. Surender Suresh Kumar, M Cummings, and Alexander Stimpson. Strengthening llm trust boundaries: a survey of prompt injection attacks. In *2024 IEEE 4th International Conference on Human-Machine Systems (ICHMS)*, pages 1–6, 2024.
16. Dong Shu, Chong Zhang, Mingyu Jin, Zihao Zhou, and Lingyao Li. Attackeval: How to evaluate the effectiveness of jailbreak attacking on large language models. *ACM SIGKDD Explorations Newsletter*, 27(1):10–19, 2025.
17. Sibo Yi, Yule Liu, Zhen Sun, Tianshuo Cong, Xinlei He, Jiaxing Song, Ke Xu, and Qi Li. Jailbreak attacks and defenses against large language models: A survey. *arXiv preprint arXiv:2407.04295*, 2024.

18. Florian Tramèr, Fan Zhang, Ari Juels, Michael K Reiter, and Thomas Ristenpart. Stealing machine learning models via prediction {APIs}. In *25th USENIX security symposium (USENIX Security 16)*, pages 601–618, 2016.
19. Kaixiang Zhao, Lincan Li, Kaize Ding, Neil Zhenqiang Gong, Yue Zhao, and Yushun Dong. A survey on model extraction attacks and defenses for large language models. In *Proceedings of the 31st ACM SIGKDD Conference on Knowledge Discovery and Data Mining V. 2*, pages 6227–6236, 2025.
20. Hongsheng Hu, Shuo Wang, Jiamin Chang, Haonan Zhong, Ruoxi Sun, Shuang Hao, Haojin Zhu, and Minhui Xue. A duty to forget, a right to be assured? exposing vulnerabilities in machine unlearning services. *arXiv preprint arXiv:2309.08230*, 2023.
21. Zirui Huang, Yunlong Mao, and Sheng Zhong. {UBA-Inf}: Unlearning activated backdoor attack with {Influence-Driven} camouflage. In *33rd USENIX Security Symposium (USENIX Security 24)*, pages 4211–4228, 2024.
22. Robert Geirhos, Jörn-Henrik Jacobsen, Claudio Michaelis, Richard Zemel, Wieland Brendel, Matthias Bethge, and Felix A Wichmann. Shortcut learning in deep neural networks. *Nature Machine Intelligence*, 2(11):665–673, 2020.
23. Ann-Kathrin Dombrowski, Maximillian Alber, Christopher Anders, Marcel Ackermann, Klaus-Robert Müller, and Pan Kessel. Explanations can be manipulated and geometry is to blame. *Advances in neural information processing systems*, 32, 2019.
24. Eslam Zaher, Maciej Trzaskowski, Quan Nguyen, and Fred Roosta. Manifold integrated gradients: Riemannian geometry for feature attribution. *arXiv preprint arXiv:2405.09800*, 2024.
25. Zhaobo Lu, Hai Liang, Minghao Zhao, Qingzhe Lv, Tiancai Liang, and Yilei Wang. Label-only membership inference attacks on machine unlearning without dependence of posteriors. *International Journal of Intelligent Systems*, 2022.
26. Ahmed Salem, Apratim Bhattacharya, Michael Backes, Mario Fritz, and Yang Zhang. Updates-leak: Data set inference and reconstruction attacks in online learning. In *29th USENIX Security Symposium, USENIX Security 2020, August 12-14, 2020*, pages 1291–1308. USENIX Association, 2020.
27. Matt Fredrikson, Somesh Jha, and Thomas Ristenpart. Model inversion attacks that exploit confidence information and basic countermeasures. In *Proceedings of the 22nd ACM SIGSAC conference on computer and communications security*, pages 1322–1333, 2015.
28. Cynthia Dwork and Aaron Roth. The algorithmic foundations of differential privacy. *Foundations and trends® in theoretical computer science*, 9(3-4):211–487, 2014.
29. Andrew C Yao. Protocols for secure computations. In *23rd annual symposium on foundations of computer science (sfcs 1982)*, pages 160–164. IEEE, 1982.
30. Jung Hee Cheon, Andrey Kim, Miran Kim, and Yongsoo Song. Homomorphic encryption for arithmetic of approximate numbers. In *International conference on the theory and application of cryptology and information security*, pages 409–437. Springer, 2017.

The manufacturer's authorised representative in the EU is Springer Nature Customer Service Centre GmbH, Europaplatz 3, 69115 Heidelberg, Germany. If you have any concerns regarding our products, please contact ProductSafety@springernature.com

Printed and bound by CPI Group (UK) Ltd, Croydon, CR0 4YY
22/09/2026
02220579-0001